THEOREMS OF LERAY-SCHAUDER TYPE
AND APPLICATIONS

SERIES IN MATHEMATICAL ANALYSIS AND APPLICATIONS

Series in Mathematical Analysis and Applications (SIMAA) is edited by Ravi P. Agarwal, National University of Singapore and Donal O'Regan, National University of Ireland.

The series is aimed at reporting on new developments in mathematical analysis and applications of a high standard and of current interest. Each volume in the series is devoted to a topic in analysis that has been applied, or is potentially applicable, to the solutions of scientific, engineering and social problems.

Volume 1
Method of Variation of Parameters for Dynamic Systems
V. Lakshmikantham and S.G. Deo

Volume 2
Integral and Integrodifferential Equations: Theory, Methods and Applications
edited by Ravi P. Agarwal and Donal O'Regan

Volume 3
Theorems of Leray-Schauder Type and Applications
Donal O'Regan and Radu Precup

THEOREMS OF LERAY-SCHAUDER TYPE AND APPLICATIONS

Donal O'Regan

Department of Mathematics
National University of Ireland
Galway, Ireland

and

Radu Precup

Department of Mathematics
"Babeş-Bolyai" University
Cluj, Romania

CRC Press
Taylor & Francis Group
Boca Raton London New York

CRC Press is an imprint of the
Taylor & Francis Group, an **informa** business

CRC Press
Taylor & Francis Group
6000 Broken Sound Parkway NW, Suite 300
Boca Raton, FL 33487-2742

© 2001 by Taylor & Francis Group, LLC
CRC Press is an imprint of Taylor & Francis Group, an Informa business

First issued in paperback 2019

No claim to original U.S. Government works

ISBN 13: 978-0-367-45472-2 (pbk)
ISBN 13: 978-90-5699-295-8 (hbk)

Visit the Taylor & Francis Web site at
http://www.taylorandfrancis.com

and the CRC Press Web site at
http://www.crcpress.com

To Alice with love
DOR

To my parents Leonida and Iordana
RP

Contents

Preface

In this book we present basic continuation theorems for several classes of nonlinear operators and typical applications to differential equations. Our approach is elementary and does not use degree theory. In addition, in this book we present in a global setting various Leray-Schauder type theorems from nonlinear analysis. Part of the material comes from the authors' own work and some of this material has been generalized here. These results together with new applications appear here for the first time. The selected topics in the book reflect the particular interests of the authors; no attempt was made to cover every area in this vast field. For example, we did not discuss topics such as Leray-Schauder type theorems for maps on locally convex spaces, A-proper maps, or set-valued maps to name but a few. The bibliography includes only referenced titles. The text is essentially self contained so it can be seen as an introduction to topological methods in nonlinear analysis. We hope it will be of interest to mathematicians, old and young, who would like to become acquainted with the rapidly progressing field of nonlinear analysis.

1. Overview

The theorems of Leray-Schauder type, also called *continuation theorems*, represent a powerful existence tool in studying operator equations and inclusions (of particular interest is the theory of nonlinear differential equations). Roughly speaking, by means of a continuation theorem we can obtain a solution of a given equation if we start from one of the solutions of a simpler equation. To be more explicit, let us consider two nonempty sets Ξ and Θ, a proper subset B of Θ and a map $\Gamma_1 : \Xi \to \Theta$. Suppose that we are interested in the solvability of the inclusion

$$\Gamma_1(x) \in B. \qquad (1.1)$$

The main idea of any continuation method for (1.1) consists in joining this inclusion to a 'simpler' one,

$$\Gamma_0(x) \in B \qquad (1.2)$$

by means of an 'homotopy' $\eta : \Xi \times [0,1] \to \Theta$ in such a way that

$$\eta(\,.\,,0) = \Gamma_0 \quad \text{and} \quad \eta(\,.\,,1) = \Gamma_1.$$

The continuation theorem contains conditions which guarantee that the solvability of (1.2) implies the solvability of (1.1). Intuitively, this occurs when one of the solutions of (1.2) can be 'continued' in a solution of

$$\eta(x,\lambda) \in B \qquad (1.3)$$

for each $\lambda \in [0,1]$, and in particular, in a solution of (1.1) for $\lambda = 1$. Such a global continuation usually follows from a local continuation in

1

Theorem 1.4 *Let* $T : \overline{U} \to X$ *be continuous and assume that for some* $x_0 \in U$, *the following condition is satisfied:*

$$S \subset \overline{U} \text{ countable, } S \subset \overline{conv}\{\{x_0\} \cup T(S)\} \implies \overline{S} \text{ compact.}$$

In addition assume that

$$(1 - \lambda) x_0 + \lambda T(x) \neq x \quad \text{for all } x \in \partial U \text{ and } \lambda \in [0, 1].$$

Then T *has at least one fixed point in* U.

We point out that each extension of Theorem 1.2 is based on a fixed point theorem for self-maps of a subset of X. Thus, Theorem 1.2 for set-contractions is derived from the Darbo fixed point principle, Theorem 1.2 for condensing maps is implied by the Sadovskii fixed point theorem, while Theorem 1.4 follows from a fixed point theorem also due to Mönch involving self-maps of a closed convex subset of X, which is a common generalization of the Schauder, Darbo and Sadovskii theorems. Therefore, we can expect that each fixed point principle for self-maps of a set yields a continuation theorem of Leray-Schauder type. This is also the case in the Browder-Göhde-Kirk fixed point principle for non-expansive self-maps of a closed bounded and convex set of a uniformly convex Banach space. The corresponding Leray-Schauder type theorem for Hilbert spaces is due to Gatica and Kirk [50]:

Theorem 1.5 *Suppose* X *is a Hilbert space,* $U \subset X$ *open bounded with* $0 \in U$ *and* $T : \overline{U} \to X$ *is nonexpansive, i.e.*

$$|T(x) - T(y)| \leq |x - y| \quad \text{for all } x, y \in \overline{U}.$$

In addition assume that the following boundary condition is satisfied:

$$\lambda T(x) \neq x \quad \text{for } x \in \partial U \text{ and } \lambda \in [0, 1].$$

Then T *has at least one fixed point in* U.

Surprisingly, one had to wait for the nineties to get a Leray-Schauder type theorem for contractions on complete metric spaces accompanying the Banach fixed point principle. The result is due to Granas [56]:

Theorem 1.6 *Let* (K, d) *be a complete metric space,* $U \subset K$ *open and* $H : \overline{U} \times [0, 1] \to K$. *Assume that the following conditions are satisfied:*

(1) *there is* $\rho \in [0,1)$ *such that*

$$d\left(H\left(x,\lambda\right), H\left(y,\lambda\right)\right) \leq \rho d\left(x, y\right)$$

for all $x, y \in \overline{U}$ *and* $\lambda \in [0,1]$;

(2) $H\left(x,\lambda\right) \neq x$ *for all* $x \in \partial U$ *and* $\lambda \in [0,1]$;

(3) $H\left(x,\lambda\right)$ *is continuous in* λ, *uniformly for* $x \in \overline{U}$.

If $H\left(.,0\right)$ *has a fixed point, then* $H\left(.,1\right)$ *also has.*

As the reader can see, there are two main approaches to the theory of the continuation methods. One uses the subtle notion of degree, while the other is based upon fixed point theory. In this book we adopt the second approach (Granas' approach) both in theory and applications. Other contributions to Granas' theory of continuation theorems and their applications have been given by Furi, Martelli, Vignoli [47], Granas, Guenther, Lee [57], [58], Krawcewicz [76], [77], Frigon [43], Lee, O'Regan [82] and Precup [134].

Organization of the book

Beginning in 1989 [130]-[132], [134]-[136], [142], the second author has developed a unified abstract theory of Granas type continuation theorems. A slight extension of this theory is presented in Chapter 9. We now state the abstract version of the topological transversality theorem.

Let A be a proper subset of Ξ, M a class of maps Γ from Ξ into Θ with $\Gamma^{-1}\left(B\right) \cap A = \emptyset$ and let ν be any map defined at least on the following class of sets $\left\{\Gamma^{-1}\left(B\right); \ \Gamma \in M\right\} \cup \left\{\emptyset\right\}$, with values in a nonempty set Z. A map $\Gamma \in M$ is said to be ν-*essential* if

$$\nu\left(\Gamma^{-1}\left(B\right)\right) = \nu\left(\Gamma'^{-1}\left(B\right)\right) \neq \nu\left(\emptyset\right)$$

for any $\Gamma' \in M$ having the same restriction to A as Γ. Also, consider an equivalence relation $\approx$ on M. The main result of Chapter 9 is the following theorem.

Theorem 1.7 *Assume that the following conditions are satisfied:*

(A) *if* $\Gamma, \Gamma' \in M$ *and* $\Gamma|_A = \Gamma'|_A$, *then* $\Gamma \approx \Gamma'$;

(H) *if* $\Gamma \approx \Gamma'$, *then there is a map* $\eta : \Xi \times [0,1] \to \Theta$ *and a function* $\upsilon : \Xi \to [0,1]$ *such that* $\eta(.,0) = \Gamma$, $\eta(.,1) = \Gamma'$, $\eta(.,\upsilon(.)) \in M$ *and*

$$\upsilon(x) = \begin{cases} 1 & \text{for } x \in \Sigma_\eta \\ 0 & \text{for } x \in A, \end{cases}$$

where $\Sigma_\eta = \{x \in \Xi; \ \eta(x,\lambda) \in B \text{ for some } \lambda \in [0,1]\}$.

Let Γ_0, $\Gamma_1 \in M$ *be two maps with* $\Gamma_0 \approx \Gamma_1$. *If* Γ_0 *is* ν-*essential, then* Γ_1 *is* ν-*essential too and*

$$\nu\left(\Gamma_0^{-1}(B)\right) = \nu\left(\Gamma_1^{-1}(B)\right).$$

Roughly speaking, the map ν measures the 'size' of some subsets of Ξ. Throughout this book, we let

$$\nu(\Psi) = \begin{cases} 1 & \text{if } \emptyset \neq \Psi \subset \Xi \\ 0 & \text{if } \Psi = \emptyset. \end{cases}$$

Theorem 1.7 makes it possible to understand globally particular continuation theorems for a great variety of single and set-valued maps in metric, locally convex or Banach spaces. Notice that this result also yields particular continuation theorems which have no corresponding degree theory analogue (see [136]).

In fact, we could start our work by the axiomatic theory presented in Chapter 9 and after that, continue with specific continuation theorems for each particular class of maps. However, in order to make the book more accessible, we prefer to discuss particular theorems successively and illustrate their applicability by means of initial value and boundary value problems for several classes of nonlinear differential equations.

In Chapter 2, we present Leray-Schauder type theorems for contractions on metric spaces. The main result is Theorem 2.3, a generalization in Maia's sense of Theorem 1.6. Applications are given for the Cauchy problem and two point boundary value problems in Banach spaces. Using Theorem 2.3 we may work in the spaces C^k endowed with L^p-norms (incomplete spaces), instead of Sobolev spaces (complete spaces).

Chapter 3 is devoted to the Leray-Schauder type theorems for nonexpansive maps. We present the extension of Theorem 1.5 to uniformly convex Banach spaces [104], [139]. As an example, we discuss Sturm-Liouville two point boundary value problems in uniformly convex Banach spaces.

Continuation theorems for accretive maps and an application to boundary value problems in Hilbert spaces are presented in Chapter 4.

In Chapter 5 we first survey Leray-Schauder type theorems for compact maps, set-contractions, condensing and Mönch type maps. Then we apply the abstract principles to the Cauchy problem and two point boundary value problems in Banach spaces. The results, Theorems 5.8, 5.9, 5.10 and 5.11 may be compared with some earlier results by Mönch [93], Mönch, von Harten [94] and Frigon, Lee [46]. For further applications and theory we refer the reader to [2]-[5], [102] and [107].

Chapter 6 is entirely devoted to applications of the fixed point principles (Banach, Schauder, Darbo and Sadovskii) and of the continuation principles for completely continuous maps and set-contractions to semilinear elliptic boundary value problems with linear growth. We discuss the existence of weak solutions under nonresonance conditions. The results and techniques may be compared with those of Mawhin, Ward Jr. [91] and Hai, Schmitt [62].

In Chapter 7 we complement the existing literature (see [158], [76], [58], [143]) with some new Leray-Schauder type theorems for semilinear operator equations of the form $Lx = T(x)$, where L is a linear Fredholm map of index zero. An application is given to periodic solutions of some first order differential systems.

In all continuation theorems presented so far, the homotopies H are defined on a set of the form $\overline{U} \times [0, 1]$, where $U \subset X$ and so, all operators $H(., \lambda)$, $\lambda \in [0, 1]$, have the same domain $\overline{U}$. In Chapter 8 we deal with the more general case when the homotopies H are defined on $\overline{\mathcal{U}}$, where $\mathcal{U} \subset X \times [0, 1]$. Notice that an analogue of Theorem 1.1 for this case is also known in Leray-Schauder degree theory. However, one had to wait for the nineties for applications. The difficulty consists in the construction of a set $\mathcal{U}$ accompanying a branch of solutions of equations with parameter, when the set of all solutions is not bounded. Beginning in 1990, Capietto, Mawhin and Zanolin [23]-[24] (see also [22], [89]) used continuation methods to discuss superlinear ordinary differential equations, in the absence of a priori bounds of solutions. The technique they provided can be viewed as an alternative to the bifurcation method (see for example [147]). Their results are based on the concept of coincidence degree, an extension of the Leray-Schauder degree. Our goal for Chapter 8 is to develop a new approach (see also [138], [142]-[143]) based on the notion of essential map, which does not make use of degree theory.

An application to periodic solutions of superlinear singular differential equations is also presented.

Chapter 10 presents Leray-Schauder theorems of cone compression and cone expansion type [61], [65]-[66], [106], [136], [156]. As an application we present a multiple solution result for higher order boundary value problems.

In Chapter 11 we state local versions (local implicit function theorems) for some of the continuation theorems presented so far and we discuss their applications to stable solutions of nonlinear problems.

Finally, we want to mention some excellent survey papers and monographs on continuation theorems and their applications: Granas, Guenther, Lee [57]-[58], Martelli [86], Mawhin [89], Mawhin, Rybakowski [90] and Zeidler [160].

2. Theorems of Leray-Schauder Type for Contractions

This chapter presents fixed point theorems of Leray-Schauder type for metric contractions. These theorems are then used to establish existence and uniqueness principles for initial value problems and two point boundary value problems in Banach or Hilbert spaces.

2.1 The Continuation Principle for Contractions on Spaces with Two Metrics

We first recall Banach's fixed point theorem (contraction principle):

Proposition 2.1 *Let* (K, d) *be a complete metric space. Suppose that* $T : K \to K$ *is a contraction, i.e. there is* $\rho \in [0, 1)$ *such that*

$$d\left(T\left(x\right), T\left(y\right)\right) \leq \rho\, d\left(x, y\right) \quad \text{for all } x, y \in K.$$

Then T *has a unique fixed point* x^* *and for any* $x \in K$, *one has*

$$d\left(T^k\left(x\right), x^*\right) \leq \frac{\rho^k}{1 - \rho}\, d\left(x, T\left(x\right)\right) \quad (k \in \mathbf{N}).$$

The following extension of Banach's fixed point theorem for contractions on spaces with two metrics is essentially due to Maia [85] (see also [126], [149] and, for some applications, [120], [121]).

9

The second proof shows that Maia's theorem in K is Banach's theorem in the completion $\widetilde{K}$, but with the fixed point in K.

The corresponding Leray-Schauder type result is the following theorem [146].

Theorem 2.3 *Let (K, d') be a complete metric space and d another metric on K. Let $D \subset K$ be d'-closed and let $U \subset K$ be d-open with $U \subset D$. Assume that for $H : D \times [0, 1] \to K$, the following conditions are satisfied:*

(i) *there is $\rho \in [0, 1)$ such that*

$$d\left(H\left(x, \lambda\right), H\left(y, \lambda\right)\right) \le \rho\, d\left(x, y\right) \qquad (2.5)$$

holds for all $x, y \in D$ and $\lambda \in [0, 1]$;

(ii) *$H\left(x, \lambda\right) \ne x$ for all $x \in D \setminus U$ and $\lambda \in [0, 1]$;*

(iii) *H is uniformly continuous from $D \times [0, 1]$ endowed with metric d on D into (K, d');*

(iv) *H is continuous from $D \times [0, 1]$ endowed with metric d' on D into (K, d');*

(v) *$H\left(x, \lambda\right)$ is continuous in λ with respect to d, uniformly for $x \in U$.*

In addition suppose that there exists a nonempty set $D_0 \subset D$ with $H_0\left(D_0\right) \subset D_0$. Then, for each $\lambda \in [0, 1]$, there exists a unique fixed point $x\left(\lambda\right)$ of H_λ. Moreover, $x\left(\lambda\right)$ depends d-continuously on λ and there exists $0 < r \le \infty$, integers $m, n_1, n_2, ..., n_{m-1}$ and numbers $0 < \lambda_1 < \lambda_2 < ... < \lambda_{m-1} < \lambda_m = 1$ such that for any $x_0 \in K$ satisfying $d\left(x_0, x\left(0\right)\right) \le r$, the sequences $(x_{j,k})_{k \ge 0}$, $j = 1, 2, ..., m$,

$$
\begin{aligned}
x_{1,0} &= x_0 \\
x_{j,k+1} &= H_{\lambda_j}(x_{j,k}), \quad k = 0, 1, ... \\
x_{j+1,0} &= x_{j,n_j}, \quad j = 1, 2, ..., m - 1
\end{aligned}
$$

are well defined and satisfy

$$d(x_{j,k}, x(\lambda_j)) \le \frac{\rho^k}{1 - \rho}\, d(x_{j,0}, H_{\lambda_j}(x_{j,0})) \quad (k \in \mathbf{N}) \qquad (2.6)$$

with

$$d'(x_{j,k}, x(\lambda_j)) \to 0 \quad as \ k \to \infty. \qquad (2.7)$$

Remark 2.1 *Obviously, we have*

$$x_{j,k} = H^k_{\lambda_j}(H^{n_{j-1}}_{\lambda_{j-1}}(\dots(H^{n_1}_{\lambda_1}(x_0))\dots)) \quad (k \in \mathbf{N}),$$

$$d(x_{j,k}, x(\lambda_j)) \to 0 \quad and \quad d'(x_{j,k}, x(\lambda_j)) \to 0$$

as $k \to \infty$ $(j = 1, 2, \dots, m)$. *In particular, for* $j = m$, $(x_{m,k})_{k \geq 0}$ *is a sequence of successive approximations of* $x(1)$, *with respect to both metrics* d *and* d'.

Proof. 1) First we prove for each $\lambda \in [0,1]$ that H_λ has a fixed point. Let

$$\Lambda_H = \{\lambda \in [0,1]; \quad H(x,\lambda) = x \quad \text{for some } x \in U\}.$$

Note $0 \in \Lambda_H$ since $H_0(D_0) \subset D_0$ and Proposition 2.2 guarantee that H_0 has a fixed point. Hence Λ_H is nonempty. We will show that Λ_H is both closed and open in $[0,1]$ and so, by the connectedness of $[0,1]$, $\Lambda_H = [0,1]$.

To prove that Λ_H is closed, let $\lambda_k \in \Lambda_H$ with $\lambda_k \to \lambda \in [0,1]$ as $k \to \infty$. Since $\lambda_k \in \Lambda_H$, there is $x_k \in U$ with $H(x_k, \lambda_k) = x_k$. Then, from (i), we obtain

$$d(x_k, x_j) = d(H(x_k, \lambda_k), H(x_j, \lambda_j)) \leq d(H(x_k, \lambda_k), H(x_k, \lambda))$$

$$+ d(H(x_k, \lambda), H(x_j, \lambda)) + d(H(x_j, \lambda), H(x_j, \lambda_j))$$

$$\leq d(H(x_k, \lambda_k), H(x_k, \lambda)) + \rho d(x_k, x_j) + d(H(x_j, \lambda), H(x_j, \lambda_j)).$$

It follows that

$$d(x_k, x_j) \leq \frac{1}{1-\rho}[d(H(x_k, \lambda_k), H(x_k, \lambda)) + d(H(x_j, \lambda), H(x_j, \lambda_j))].$$

This, with (v), shows that the sequence (x_k) is d-Cauchy. Furthermore, from $d'(x_k, x_j) = d'(H(x_k, \lambda_k), H(x_j, \lambda_j))$ and (iii), we see that (x_k) is also d'-Cauchy. Thus, by the completeness of d', there is an $x \in K$ with $d'(x_k, x) \to 0$ as $k \to \infty$. Since $x_k \in D$ and D is d'-closed, we have $x \in D$. As a result we have

$$d'(x_k, H(x, \lambda)) \to d'(x, H(x, \lambda)) \quad \text{as } k \to \infty$$

and, from (iv), we also have

$$d'(x_k, H(x, \lambda)) = d'(H(x_k, \lambda_k), H(x, \lambda)) \to 0 \quad \text{as } k \to \infty.$$

Hence $d'(x, H(x, \lambda)) = 0$, that is $H(x, \lambda) = x$. From (ii), $x \in U$ and so $\lambda \in \Lambda_H$.

To prove that Λ_H is open in $[0, 1]$, let $\mu \in \Lambda_H$ and $z \in U$ with $H(z, \mu) = z$. Since U is d-open, there exists a closed d-ball $B = \overline{B}_\delta(z)$, $\delta > 0$, with $B \subset U$. Notice (v) guarantees that there is $\eta = \eta(\delta) > 0$ with

$$d(z, H(z, \lambda)) = d(H(z, \mu), H(z, \lambda)) \leq (1 - \rho)\delta \qquad (2.8)$$

for $|\lambda - \mu| \leq \eta$. Consequently,

$$d(z, H(x, \lambda)) \leq d(z, H(z, \lambda)) + d(H(z, \lambda), H(x, \lambda))$$

$$\leq (1 - \rho)\delta + \rho d(z, x) \leq \delta$$

whenever $x \in B$ and $|\lambda - \mu| \leq \eta$. This shows that for $|\lambda - \mu| \leq \eta$, H_λ sends B into itself. Let B' be the d'-closure of B. It is easily seen that $H_\lambda(B') \subset B'$ for $|\lambda - \mu| \leq \eta$. Now we may apply Proposition 2.2 to $T = H_\lambda$. Consequently, there exists a $x(\lambda) \in B' \subset D$ a fixed point of H_λ for $|\lambda - \mu| \leq \eta$. This shows that μ is an interior point of Λ_H. Hence Λ_H is open in $[0, 1]$. Note from Proposition 2.2, that for every $x \in B$ and $|\lambda - \mu| \leq \eta$, we have that the sequence $(H_\lambda^k(x))_{k \geq 0}$ is well defined,

$$d(H_\lambda^k(x), x(\lambda)) \leq \frac{\rho^k}{1 - \rho} d(x, H_\lambda(x)) \quad (k \in \mathbf{N})$$

and

$$d'(H_\lambda^k(x), x(\lambda)) \to 0 \quad \text{as } k \to \infty.$$

2) The uniqueness of $x(\lambda)$ is a simple consequence of (i).
3) $x(\lambda)$ is d-continuous on $[0, 1]$. Indeed,

$$d(x(\lambda), x(\mu)) = d(H(x(\lambda), \lambda), H(x(\mu), \mu))$$

$$\leq d(H(x(\lambda), \lambda), H(x(\mu), \lambda)) + d(H(x(\mu), \lambda), H(x(\mu), \mu))$$

$$\leq \rho d(x(\lambda), x(\mu)) + d(H(x(\mu), \lambda), H(x(\mu), \mu)).$$

This, with (v), implies

$$d(x(\lambda), x(\mu)) \leq \frac{1}{1 - \rho} d(H(x(\mu), \lambda), H(x(\mu), \mu)) \to 0$$

as $\lambda \to \mu$.

4) Finding r. For any $\mu \in [0,1]$, let

$$r(\mu) = \inf \{d(x, x(\mu)); \ x \in K \setminus U\}.$$

Since $x(\mu) \in U$ and U is d-open, $r(\mu) > 0$. We claim that

$$\inf \{r(\mu); \ \mu \in [0,1]\} > 0. \tag{2.9}$$

To prove this, assume the contrary. Then, there are $\mu_k \in [0,1]$ with $r(\mu_k) \to 0$ as $k \to \infty$. Clearly, we may assume that $\mu_k \to \mu$ for some $\mu \in [0,1]$. From the d-continuity of $x(\lambda)$ we have

$$d(x(\mu_k), x(\mu)) < r(\mu)/2 \quad \text{for } k \geq k_1. \tag{2.10}$$

On the other hand, since $r(\mu_k) \to 0$, we have

$$r(\mu_k) < r(\mu)/2 \quad \text{for } k \geq k_2. \tag{2.11}$$

Let $k_0 = \max\{k_1, k_2\}$. Notice (2.11)' and the definition of $r(\mu_{k_0})$ as infimum, guarantee that there is an $x \in K \setminus U$ with

$$d(x, x(\mu_{k_0})) < r(\mu)/2. \tag{2.12}$$

Then, using (2.10) and (2.12), we obtain

$$\begin{aligned} d(x, x(\mu)) &\leq d(x, x(\mu_{k_0})) + d(x(\mu_{k_0}), x(\mu)) \\ &< 2r(\mu)/2 = r(\mu), \end{aligned}$$

a contradiction. Thus (2.9) holds as claimed. Now we choose any $r > 0$ less than the infimum in (2.9), with the convention that $r = \infty$ if the infimum equals infinity.

5) Finding m **and** $0 < \lambda_1 < \lambda_2 < \ldots < \lambda_{m-1} < 1$. Let $h = \eta(r)$, where r has been fixed in the previous step and $\eta(r)$ is chosen as in (2.8). Then, by what was shown at the end of step 1), for each $\mu \in [0,1]$,

$$\begin{aligned} d(x, x(\mu)) &\leq r \quad \text{and} \quad |\lambda - \mu| \leq h \quad \text{imply} \\ &(H_\lambda^k(x))_{k \geq 0} \quad \text{is well defined,} \end{aligned} \tag{2.13}$$

$$d(H_\lambda^k(x), x(\lambda)) \leq \frac{\rho^k}{1-\rho} d(x, H_\lambda(x)) \quad (k \in \mathbf{N}) \text{ and}$$

$$d'(H_\lambda^k(x), x(\lambda)) \to 0 \quad \text{as } k \to \infty.$$

Now we choose any partition

$$0 = \lambda_0 < \lambda_1 < \dots < \lambda_{m-1} < \lambda_m = 1$$

of $[0, 1]$ such that $\lambda_{j+1} - \lambda_j \le h$, $j = 0, 1, \dots, m-1$.

6) Finding the integers $n_1, n_2, \dots, n_{m-1}$. Now

$$d(x_{1,0}, x(0)) = d(x_0, x(0)) \le r, \quad \lambda_1 - \lambda_0 \le h,$$

and (2.13) guarantees that $(x_{1,k})_{k \ge 0}$ is well defined and satisfies (2.6)-(2.7). From (2.6) we may choose $n_1 \in \mathbf{N}$ with $d(x_{1,n_1}, x(\lambda_1)) \le r$. Now, we have

$$d(x_{2,0}, x(\lambda_1)) = d(x_{1,n_1}, x(\lambda_1)) \le r \quad \text{and} \quad \lambda_2 - \lambda_1 \le h.$$

From the above argument, $(x_{2,k})_{k \ge 0}$ is well defined and satisfies (2.6)-(2.7). In general, at step j $(1 \le j \le m-1)$, we choose $n_j \in \mathbf{N}$ such that $d(x_{j,n_j}, x(\lambda_j)) \le r$. Then

$$d(x_{j+1,0}, x(\lambda_j)) = d(x_{j,n_j}, x(\lambda_j)) \le r, \quad \lambda_{j+1} - \lambda_j \le h,$$

and (2.13) guarantees that the sequence $(x_{j+1,k})_{k \ge 0}$ is well defined and satisfies (2.6)-(2.7). $\square$

The above proof yields the following algorithm for the approximation of $x(1)$ under the assumptions of Theorem 2.3:

Suppose we know r and h and we wish to obtain an approximation $\bar{x}_1$ of $x(1)$ with $d(\bar{x}_1, x(1)) \le \varepsilon$. Then we choose any partition $0 = \lambda_0 < \lambda_1 < \lambda_2 < \dots < \lambda_{m-1} < \lambda_m = 1$ of $[0, 1]$ with $\lambda_{j+1} - \lambda_j \le h$, $j = 0, 1, \dots, m-1$, and any element x_0 with $d(x_0, x(0)) \le r$. Now follow an iterative procedure.

Iterative procedure:

Set $n_0 := 0$ and $x_{0,n_0} := x_0$;
For $j := 1$ to $m-1$ do
$\qquad x_{j,0} := x_{j-1,n_{j-1}}$
$\qquad k := 0$
$\qquad$ while $\rho^k (1-\rho)^{-1} d(x_{j,0}, H_{\lambda_j}(x_{j,0})) > r$ do
$\qquad\qquad x_{j,k+1} := H_{\lambda_j}(x_{j,k})$
$\qquad\qquad k := k+1$
$\qquad n_j = k$

Set $k := 0$
While $\rho^k (1 - \rho)^{-1} d(x_{m,0}, H_1(x_{m,0})) > \varepsilon$ do
$\qquad x_{m,k+1} = H_1(x_{m,k})$
$\qquad k := k + 1$
Finally take $\bar{x}_1 = x_{m,k}$.

Remark 2.2 *Clearly, if $d \le d'$ on K, then it suffices that the estimates in the above algorithm be made with respect to d' instead of d.*

Notice that, when $D = U = K$ and $H_\lambda = T$ for all $\lambda \in [0,1]$, Theorem 2.3 reduces to Proposition 2.2. In this case, $r = \infty$ and $m = 1$.

In case that $d = d'$, Theorem 2.3 yields the following computational version of Granas continuation principle for contractions on complete metric spaces (Theorem 1.6).

Theorem 2.4 *Let (K, d) be a complete metric space, $U \subset K$ open and $H : \bar{U} \times [0,1] \to K$. Assume that the following conditions are satisfied:*

(a1) *there is $\rho \in [0,1)$ such that*

$$d(H(x, \lambda), H(y, \lambda)) \le \rho \, d(x, y)$$

whenever $x, y \in \bar{U}$ and $\lambda \in [0,1]$;

(a2) $H(x, \lambda) \ne x$ *for all $x \in \partial U$ and $\lambda \in [0,1]$;*

(a3) $H(x, \lambda)$ *is continuous in λ, uniformly for $x \in \bar{U}$.*

In addition suppose that there exists a nonempty set $D_0 \subset \bar{U}$ with $H_0(D_0) \subset D_0$. Then, for each $\lambda \in [0,1]$, there exists a unique fixed point $x(\lambda)$ of H_λ. Moreover, $x(\lambda)$ depends continuously on λ and there exists $0 < r \le \infty$, integers m, n_1, n_2, ..., n_{m-1} and numbers $0 < \lambda_1 < \lambda_2 < ... < \lambda_{m-1} < \lambda_m = 1$ such that for any $x_0 \in K$ satisfying $d(x_0, x(0)) \le r$, the sequences $(x_{j,k})_{k \ge 0}$, $j = 1, 2, ..., m$,

$$\begin{aligned}
x_{1,0} &= x_0 \\
x_{j,k+1} &= H_{\lambda_j}(x_{j,k}), \quad k = 0, 1, ... \\
x_{j+1,0} &= x_{j,n_j}, \quad j = 1, 2, ..., m-1
\end{aligned}$$

are well defined and satisfy

$$d(x_{j,k}, x(\lambda_j)) \le \frac{\rho^k}{1 - \rho} d(x_{j,0}, H_{\lambda_j}(x_{j,0})) \quad (k \in \mathbf{N}).$$

Obviously, for $U = K$ and $H_\lambda = T$ for all $\lambda \in [0,1]$, Theorem 2.4 reduces to the Banach contraction principle.

Continuation theorems for generalized contractions can be found in [44] (for set-valued contractions), [45], [104] (for ϕ-contractions), [140] (for maps of Caristi type).

2.2 Global Solutions to the Cauchy Problem on a Bounded Set in Banach Spaces

Throughout this book we shall use the following standard notation. We denote by E any real Banach space and by $|\,.\,|$ its norm. For a real number $R > 0$ and any $u_0 \in E$, $\overline{B}_R(u_0; E)$ is the closed ball $\{u \in E;\ |u - u_0| \le R\}$. When no confusion is possible, we simply denote it by $\overline{B}_R(u_0)$.

Let $I = [0,1]$. We consider the space $C(I; E)$ of all continuous functions $u : I \to E$, endowed with the max-norm

$$\|u\|_\infty = \max_{t \in I} |u(t)|$$

and for every integer $k \ge 1$, the space $C^k(I; E)$ of all functions $u : I \to E$ such that for each $j = 1,\ 2,\ \dots,\ k$, the j-th derivative $u^{(j)}$ exists and is continuous, endowed with the max-norm

$$\|u\|_{k,\infty} = \max\left\{ \|u\|_\infty,\ \|u'\|_\infty,\ \dots,\ \|u^{(k)}\|_\infty \right\}.$$

Also, for any real $1 \le p \le \infty$, we consider the Banach space $L^p(I; E)$ of all measurable functions $u : I \to E$ such that $|u|^p$ is Lebesgue integrable on I, with the norm

$$\|u\|_p = \left(\int_0^1 |u(t)|^p\, dt \right)^{1/p}$$

for $p < \infty$ and

$$\|u\|_\infty = \inf \{ M \ge 0;\ |u(t)| \le M \ \text{a.e.} \ t \in I \}.$$

If $(E, \langle\,.\,,\,.\,\rangle)$ is a Hilbert space and $p = 2$, then $L^2(I; E)$ is a Hilbert space with inner-product

$$\langle u, v \rangle_2 = \int_0^1 \langle u(t), v(t) \rangle\, dt.$$

For any real number $1 \leq p \leq \infty$, we define the *Sobolev spaces* $W^{m,p}(I; E)$ inductively as follows. A function u belongs to $W^{1,p}(I; E)$ if it is continuous and there exists $v \in L^p(I; E)$ such that

$$u(t) = u(0) + \int_0^t v(s)\, ds, \quad t \in I,$$

where by $\int_0^t v(s)\, ds$ we mean the Bochner integral (recall that each L^p function is Bochner integrable). It is clear that if $u \in W^{1,p}(I; E)$, then u is absolutely continuous on I, is differentiable almost everywhere on I, $u' \in L^p(I; E)$, and

$$u(t) = u(0) + \int_0^t u'(s)\, ds, \quad t \in I.$$

Then for any integer $m > 1$, $u \in W^{m,p}(I; E)$ if $u, u' \in W^{m-1,p}(I; E)$.

If $u \in W^{m,p}(I; E)$, then $u \in C^{m-1}(I; E)$, $u^{(m-1)}$ is absolutely continuous on I, differentiable almost everywhere on I,

$$u^{(m)} := \left(u^{(m-1)} \right)' \in L^p(I; E),$$

and

$$u^{(m-1)}(t) = u^{(m-1)}(0) + \int_0^t u^{(m)}(s)\, ds, \quad t \in I.$$

Recall that if E is reflexive, then any absolutely continuous function $u : I \to E$ is differentiable almost everywhere on I and $u' \in L^1(I; E)$ (see [[12], Theorem 1.2.1])

The space $W^{m,p}(I; E)$ is a Banach space with the norm

$$\|u\|_{m,p} = \max \left\{ \left\| u^{(j)} \right\|_p ; \quad j = 0, 1, \ldots, m \right\}.$$

Let $W_0^{1,p}(I; E)$ denote the space of all functions $u \in W^{1,p}(I; E)$ with $u(0) = u(1) = 0$. $W_0^{1,p}(I; E)$ is a Banach space with the norm defined by

$$\|u\|_{0,p} = \|u'\|_p.$$

If $(E, \langle \, . \, , \, . \, \rangle)$ is a Hilbert space, then $W_0^{1,2}(I; E)$ is a Hilbert space with the inner-product

$$\langle u, v \rangle_{0,2} = \langle u', v' \rangle_2 = \int_0^1 \langle u'(t), v'(t) \rangle\, dt.$$

We shall also use the following notion in this book: a function f : $I \times D \to E$, where $D \subset E^n$, is said to be L^p-*Carathéodory*, if $f(\,.\,,x)$: $I \to E$ is measurable for each fixed $x \in D$, $f(t,\,.\,) : D \to E$ is continuous for a.e. $t \in I$, and for each $r > 0$, there exists $h_r \in L^p(I; \mathbf{R})$ such that

$$|f(t, x_1, x_2, ..., x_n)| \leq h_r(t)$$

for a.e. $t \in I$ and all $x = (x_1, x_2, ..., x_n) \in D$ with $|x_j| \leq r$, $j = 1, 2, ..., n$. The function f is said to be a *Carathéodory function* if it is L^1-Carathéodory.

We first describe a typical application of Theorem 2.4.

Let us consider the initial value (Cauchy) problem

$$\begin{cases} u' = f(t, u), & t \in I \\ u(0) = u_0 \end{cases} \tag{2.14}$$

in the Banach space $(E, |\,.\,|)$, where $u_0 \in E$. We seek *classical solutions* $u \in C^1(I; E)$ if f is continuous and *Carathéodory solutions* $u \in W^{1,1}(I; E)$ if f is a Carathéodory function.

If $f : I \times E \to E$ is continuous (resp., Carathéodory), then a function u is a classical (resp., Carathéodory) solution to (2.14) if and only if $u \in C(I; E)$ solves the Volterra integral equation

$$u(t) = u_0 + \int_0^t f(s, u(s))\, ds, \quad t \in I,$$

that is, u solves the fixed point problem

$$u = T(u)$$

for the integral operator $T : C(I; E) \to C(I; E)$,

$$T(u)(t) = u_0 + \int_0^t f(s, u(s))\, ds.$$

It is well known that if $f(t, u)$ satisfies a global Lipschitz condition in u, on the entire space E, then (2.14) has a unique *global solution* (that is, a solution defined on the entire interval I) which can be obtained by means of the Banach contraction principle, while if f only satisfies a local Lipschitz condition, say

$$|f(t, u) - f(t, v)| \leq l\,|u - v| \quad \text{for } |u - u_0|,\ |v - u_0| \leq R \tag{2.15}$$

and a.e. $t \in I$, then, by Banach's theorem we can only prove the existence and uniqueness of a *local solution* (that is, a solution defined on a subinterval I_R of I). The subinterval I_R comes from the invariance condition $T(\overline{B}_R(u_0)) \subset \overline{B}_R(u_0)$, where $\overline{B}_R(u_0)$ is a ball in $C(I_R; E)$. We can avoid the invariance condition and find a global solution if we use a Leray-Schauder type theorem for contractions instead of Banach's fixed point theorem.

Thus, we have the following existence principle.

Theorem 2.5 *Let* $R > 0$ *and* $f : I \times \overline{B}_R(u_0; E) \to E$. *Assume that the following conditions are satisfied:*

(h1) *f is continuous (resp., $f(., u)$ is measurable for each $u \in \overline{B}_R(u_0; E)$ and $f(., u_0) \in L^1(I; E)$);*

(h2) *there exists $l \in L^1(I; \mathbf{R})$ such that*

$$|f(t, u) - f(t, v)| \leq l(t)|u - v|$$

for a.e. $t \in I$ and all $u, v \in \overline{B}_R(u_0; E)$;

(h3) *for any solution of*

$$\begin{cases} u' = \lambda f(t, u), & t \in I \\ u(0) = u_0 \end{cases} \tag{2.16}$$

where $\lambda \in [0, 1]$, one has $|u(t) - u_0| < R$ for all $t \in I$.

Then, (2.14) has a unique classical (resp., Carathéodory) solution with $|u(t) - u_0| < R$, $t \in I$.

Proof. First we note that from (h1) and (h2), it follows that

$$|f(t, u)| \leq l(t)|u - u_0| + |f(t, u_0)| \leq R l(t) + |f(t, u_0)|,$$

where $Rl(.) + |f(., u_0)| \in L^1(I; \mathbf{R})$. This shows that f is a Carathéodory function. We apply Theorem 2.4 with $K = C(I; E)$, $U = \{u \in K; |u(t) - u_0| < R$ for all $t \in I\}$ and

$$H(u, \lambda) = (1 - \lambda) u_0 + \lambda T(u).$$

We take as d the corresponding metric to the equivalent norm of Bielecki type on $C(I; E)$,

$$\|u\|_\theta = \max_{t \in I} \left\{ |u(t)| \exp\left(-\theta \int_0^t l(\tau)\, d\tau \right) \right\},$$

for some $\theta > 1$. From (h3), we trivially have (a2), while (a3) is immediate. Now we show that (a1) also holds since $\theta > 1$. Indeed

$$|H(u, \lambda)(t) - H(v, \lambda)(t)| \leq \lambda \int_0^t |f(s, u(s)) - f(s, v(s))|\, ds$$

$$\leq \int_0^t l(s) |u(s) - v(s)|\, ds$$

$$= \int_0^t l(s) \exp\left(-\theta \int_0^s l(\tau)\, d\tau \right) |u(s) - v(s)| \exp\left(\theta \int_0^s l(\tau)\, d\tau \right) ds$$

$$\leq \|u - v\|_\theta \int_0^t l(s) \exp\left(\theta \int_0^s l(\tau)\, d\tau \right) ds$$

$$\leq \frac{1}{\theta} \|u - v\|_\theta \exp\left(\theta \int_0^t l(\tau)\, d\tau \right).$$

Consequently,

$$\|H(u, \lambda) - H(v, \lambda)\|_\theta \leq \frac{1}{\theta} \|u - v\|_\theta.$$

Now the conclusion follows since $H_0 \equiv u_0$. $\square$

To satisfy (h3) and so to obtain global solutions, we have to impose appropriate growth restrictions on f. The following existence theorem is based on a growth restriction of Wintner type (see [159]).

Theorem 2.6 *Let $R > 0$ and $f : I \times \overline{B}_R(u_0; E) \to E$. Assume that the following conditions are satisfied:*

(i) *f is continuous (resp., $f(.,u)$ is measurable for each $u \in \overline{B}_R(u_0; E)$);*

(ii) *there exists $l \in L^1(I; \mathbf{R})$ such that*

$$|f(t, u) - f(t, v)| \leq l(t) |u - v|,$$

for a.e. $t \in I$ and all $u, v \in \overline{B}_R(u_0; E)$;

(iii) *there exists* $\beta \in L^1(I; \mathbf{R})$ *and* $\psi : [0, R] \to (0, \infty)$ *nondecreasing*
   with $1/\psi \in L^1([0, R]; \mathbf{R})$ *such that*

$$|f(t, u)| \le \beta(t) \psi(|u - u_0|) \tag{2.17}$$

for a.e $t \in I, |u - u_0| \le R,$ *and*

$$\int_0^R \frac{1}{\psi(\tau)} \, d\tau > \int_0^1 \beta(\tau) \, d\tau. \tag{2.18}$$

Then (2.14) has a unique classical (resp., Carathéodory) solution
with $|u(t) - u_0| < R, \ t \in I.$

Proof. We show that condition (h3) is satisfied. Let u be any solution
to (2.16) for some $\lambda \in (0, 1]$. Then

$$u(t) = u_0 + \lambda \int_0^t f(s, u(s)) \, ds, \quad t \in I$$

and so

$$|u(t) - u_0| \le \lambda \int_0^t |f(s, u(s))| \, ds = \int_0^t |u'(s)| \, ds.$$

Let

$$w(t) = \int_0^t |u'(s)| \, ds.$$

We have $w \in W^{1,1}(I; \mathbf{R}_+)$ and $|u(t) - u_0| \le w(t)$ on I. We claim
that $w(t) < R$ for all $t \in I$. To prove this, we assume the contrary.
Then, since $w(0) = 0$, there exists a smallest $t_0 \in (0, 1]$ with $w(t) < R$
on $[0, t_0)$ and $w(t_0) = R$. Using (2.17) we obtain

$$w'(t) = |u'(t)| \le \beta(t) \psi(|u(t) - u_0|) \le \beta(t) \psi(w(t)) \tag{2.19}$$

on $[0, t_0]$. Consequently

$$\int_0^{t_0} \frac{w'(\tau)}{\psi(w(\tau))} \, d\tau \le \int_0^{t_0} \beta(\tau) \, d\tau \le \int_0^1 \beta(\tau) \, d\tau.$$

Now since

$$\int_0^{t_0} \frac{w'(\tau)}{\psi(w(\tau))} \, d\tau = \int_0^R \frac{1}{\psi(\tau)} \, d\tau,$$

we have

$$\int_0^R \frac{1}{\psi(\tau)} \, d\tau \le \int_0^1 \beta(\tau) \, d\tau,$$

a contradiction. Hence $w(t) < R$ for all $t \in I$, as claimed. $\square$

Remark 2.3 *In a Hilbert space* $(E, \langle ., . \rangle)$, *the result of Theorem 2.6 remains true if instead of condition (iii) we require*

(iii') *there exists* $\beta \in L^1(I; \mathbf{R}_+)$ *and* $\varphi : [0, R] \rightarrow (0, \infty)$ *with* $t/\varphi(t) \in L^1([0, R]; \mathbf{R})$, *such that*

$$\langle u - u_0, \, f(t, u) \rangle \leq \beta(t) \varphi(|u - u_0|)$$

for a.e. $t \in I$, $|u - u_0| \leq R$, *and*

$$\int_0^R \frac{\tau}{\varphi(\tau)} \, d\tau > \int_0^1 \beta(\tau) \, d\tau.$$

In this case we set $w(t) = |u(t) - u_0|$ *and we use*

$$w(t) \, w'(t) \; = \; \langle u(t) - u_0, \, u'(t) \rangle.$$

Then (2.19) holds with $\psi(t) = \varphi(t)/t$. *This shows that if* E *is a Hilbert space, then the assumption in Theorem 2.6 that* ψ *be nondecreasing may be removed (take* $\varphi(t) = t\psi(t)$).

Example. Let $g \in L^1(I; \mathbf{R})$, $p > 0$ and $u_0 \in E$. There exists a unique solution to

$$\begin{cases} u' = g(t) |u|^p u, & t \in I \\ u(0) = u_0 \end{cases}$$

provided that

$$|u_0|^p < 1 / \left(p \, \|g\|_{L^1(I)} \right). \tag{2.20}$$

Moreover, the result is the best possible.

Indeed, here $f(t, u) = g(t) |u|^p u$ is locally Lipschitz in u on $\mathbf{R}$. Thus condition (ii) in Theorem 2.6 holds for any $R > 0$. On the other hand,

$$|f(t, u)| \leq |g(t)| \, |u|^{p+1} \leq |g(t)| \, (|u - u_0| + |u_0|)^{p+1},$$

so (iii) holds with

$$\beta(t) \; = \; |g(t)| \quad \text{and} \quad \psi(t) = (t + |u_0|)^{p+1},$$

provided that

$$\frac{1}{p} \left[\frac{1}{|u_0|^p} - \frac{1}{(R + |u_0|)^p} \right] > \int_0^1 |g(\tau)| \, d\tau.$$

Now (2.20) guarantees that this inequality is satisfied if we choose R sufficiently large. Finally, direct integration shows that there is no solution for $E = \mathbf{R}$, $g(t) \equiv 1$, $p = 2$ and $|u_0|^2 \geq 1/2$ and so the result is the best possible.

2.3 Boundary Value Problems on a Bounded Set in Banach Spaces

We now describe a typical example of an application of Theorem 2.3.

Let us consider the two point boundary value problem

$$\begin{cases} u'' = f(t, u, u'), & t \in I = [0, 1] \\ V_1(u) = b_1, & V_2(u) = b_2 \end{cases} \tag{2.21}$$

in a Banach space $(E, |\,.\,|)$, where $b_1, b_2 \in E$, V_1, V_2 are linear continuous from $C^1(I; E)$ into E, with f defined on a bounded set of $I \times E^2$.

We let

$$C^1_{B_0} = \left\{ u \in C^1(I; E); \quad V_j(u) = 0, \ j = 1, 2 \right\}$$

and $C^k_{B_0} = C^1_{B_0} \cap C^k(I; E)$ $(k \geq 2)$. Similarly,

$$C^1_{B} = \left\{ u \in C^1(I; E); \quad V_j(u) = b_j, \ j = 1, 2 \right\}$$

and $C^k_{B} = C^1_{B} \cap C^k(I; E)$. Also, for an integer $m \geq 2$ and a real $1 \leq p \leq \infty$, we let

$$W^{m,p}_{B_0} = C^1_{B_0} \cap W^{m,p}(I; E), \quad W^{m,p}_{B} = C^1_{B} \cap W^{m,p}(I; E).$$

Recall that $W^{m,p}(I; E) \subset C^{m-1}(I; E)$.

In what follows we assume that the unique solution of $u'' = 0$ which satisfies $V_j(u) = 0$, $j = 1, 2$, is the null function. Then, there is a unique solution to

$$\begin{cases} u'' = 0, & t \in I \\ V_1(u) = b_1, & V_2(u) = b_2, \end{cases} \tag{2.22}$$

say $u_0(t)$, and there is a Green's function $G(t,s)$ corresponding to operator u'' and boundary conditions $V_j(u) = 0$, $j = 1, 2$. Moreover, for each $p \in [1, \infty]$, the operator

$$L : W_{B_0}^{2,p} \to L^p(I; E), \quad Lu = u''$$

is invertible and

$$L^{-1}v(t) = \int_0^1 G(t,s)\, v(s)\, ds.$$

The same is true for the operator

$$L : C_{B_0}^2 \to C(I; E), \quad Lu = u''.$$

Now we state a very general existence and uniqueness principle in a ball of C_B^1.

Theorem 2.7 *Let $R > 0$, $1 < p \le \infty$ and $T : D_R \to W_B^{2,p}$ be any map, where $D_R = \{u \in C_B^1; \ \|u\|_{1,\infty} \le R\}$. Assume that $\|u_0\|_{1,\infty} < R$ and that the following conditions are satisfied:*

(H1) *$T(D_R)$ is bounded in $(C^1(I; E), \|\cdot\|_{1,\infty})$ and there is $R' > 0$ such that $|u''(t)| \le R'$ for a.e. $t \in I$ and any $u \in T(D_R)$;*

(H2) *there exists a metric d on C_B^1 equivalent to the metric induced by $\|\cdot\|_{1,p}$ satisfying*

$$d(u, v) \le c_0 \|u - v\|_{1,p}$$

for all $u, v \in C_B^1$ and some $c_0 > 0$, such that

$$\|T(u) - T(v)\|_{1,\infty} \le c\, d(u, v) \tag{2.23}$$

and

$$d(T(u), T(v)) \le \rho\, d(u, v) \tag{2.24}$$

for all $u, v \in D_R$ and some $c > 0$, $\rho \in [0, 1)$;

(H3) *if $u \in D_R$ solves $u = (1 - \lambda) u_0 + \lambda T(u)$ for some $\lambda \in [0, 1]$, then $\|u\|_{1,\infty} < R$.*

Then T has a unique fixed point in D_R.

Proof. We shall apply Theorem 2.3. Denote by d' the metric induced by $\|\cdot\|_{1,\infty}$ on C_B^1. Recall that V_j, $j = 1, 2$, were supposed continuous; consequently, (C_B^1, d') is a complete metric space. Let

$$K_0 = \text{conv}\,\{\{u_0\} \cup T(D_R)\},$$

where "conv" stands for the convex hull. Since $u_0 \in C_B^1$, $T(D_R) \subset C_B^1$ and C_B^1 is convex, we also have $K_0 \subset C_B^1$. Denote by K the d'-closure of K_0 in C_B^1 and let $D = K \cap D_R$. Obviously, D is d'-closed in K.

From (H1), we see that any function u in K_0 satisfies $|u''(t)| \leq R'$ for a.e. $t \in I$. This property is the reason for this choice of K.

Define $H : D \times [0, 1] \to K$ by

$$H(u, \lambda) = (1 - \lambda)\, u_0 + \lambda T(u).$$

We now check that all the assumptions of Theorem 2.3 are satisfied, where U is the d-interior of D in K.

Condition (i) follows from (2.24) since $D \subset D_R$. From (2.23), since $T(D_R)$ is bounded in $C^1(I; E)$, we have

$$\|H(u, \lambda) - H(v, \mu)\|_{1,\infty} \tag{2.25}$$

$$\leq \|H(u, \lambda) - H(v, \lambda)\|_{1,\infty} + \|H(v, \lambda) - H(v, \mu)\|_{1,\infty}$$

$$\leq \|T(u) - T(v)\|_{1,\infty} + c'\,|\lambda - \mu| \leq c\,d(u, v) + c'\,|\lambda - \mu|,$$

where c' is a constant depending only on R. It follows that H is uniformly (d, d')-continuous, that is (iii) is true. Since d is equivalent to $\|\cdot\|_{1,p}$ and $\|\cdot\|_{1,p} \leq \|\cdot\|_{1,\infty}$, then (2.25) guarantees that (iv) is true. Now, if in (2.25) we put $u = v$, then we obtain

$$d(H(u, \lambda), H(u, \mu)) \leq c_0\,\|H(u, \lambda) - H(u, \mu)\|_{1,p}$$

$$\leq c_0\,\|H(u, \lambda) - H(u, \mu)\|_{1,\infty} \leq c_0 c'\,|\lambda - \mu|.$$

This proves (v).

It is clear that (ii) follows from (H3) if we prove that

$$u \in D \quad \text{and} \quad \|u\|_{1,\infty} < R \quad \text{implies} \quad u \in U. \tag{2.26}$$

So let $u \in D$ with $\|u\|_{1,\infty} < R$. We have to show that there exists $r > 0$ such that

$$v \in K \quad \text{and} \quad \|v - u\|_{1,p} < r \quad \text{imply} \quad v \in D_R.$$

Suppose the contrary. Then, there is a sequence $(u_k) \subset K$ with

$$\|u_k - u\|_{1,p} < 1/k \quad \text{and} \quad u_k \notin D_R.$$

Then, $|u_k(t)| > R$ or $|u'_k(t)| > R$ for some $t \in I$. On the other hand, if we let $R_0 = \|u\|_{1,\infty}$, then $R_0 < R$ and $|u(t)| \leq R_0$, $|u'(t)| \leq R_0$ for all $t \in I$. Consequently, for each k there is at least one t such that:

1) $|u_k(t) - u(t)| \geq |u_k(t)| - |u(t)| \geq |u_k(t)| - R_0 > R - R_0,$

or

2) $|u'_k(t) - u'(t)| \geq |u'_k(t)| - |u'(t)| \geq |u'_k(t)| - R_0 > R - R_0.$

We shall derive a contradiction by using the following result.

Lemma 2.8 Let $\chi \in W^{1,\infty}(I; E)$. If $|\chi(t)| \geq a > 0$ for some $t \in I$ and $|\chi'(t)| \leq M$ for a.e. $t \in I$, then

$$\int_0^1 |\chi(s)|\, ds \geq \min\left\{a/2, 3a^2/(8M)\right\}.$$

I. First suppose that $|u'_k(t) - u'(t)| \leq R - R_0$ for all $t \in I$ and for infinitely many values of k. Then, passing if necessary to a subsequence, we may assume that for any k, we have

$$|u'_k(t) - u'(t)| \leq R - R_0 \quad \text{for all } t \in I \text{ and}$$

$$|u_k(t) - u(t)| > R - R_0 \quad \text{for at least one } t.$$

Then, by Lemma 2.8, it follows that

$$\int_0^1 |u_k(s) - u(s)|\, ds \geq 3(R - R_0)/8 > 0$$

for all k. This implies $\|u_k - u\|_{1,p} \not\to 0$ as $k \to \infty$, a contradiction.

II. In the opposite case to I, we may suppose that for any k, we have

$$|u'_k(t) - u'(t)| > R - R_0 \quad \text{for at least one } t.$$

Let $\varepsilon > 0$. Since $u, u_k \in K$, there are $\tilde{u}, \tilde{u}_k \in K_0$ with

$$|\tilde{u}'_k(t) - \tilde{u}'(t)| > R - R_0 \quad \text{for at least one } t,$$

$$\int_0^1 |u'_k(s) - \tilde{u}'_k(s)|\, ds \leq \varepsilon/2 \quad \text{and} \quad \int_0^1 |u'(s) - \tilde{u}'(s)|\, ds \leq \varepsilon/2.$$

Also since $\tilde{u}, \tilde{u}_k \in K_0$, we have

$$|\tilde{u}_k''(t) - \tilde{u}''(t)| \leq |\tilde{u}_k''(t)| + |\tilde{u}''(t)| \leq 2R' \quad \text{for a.e. } t \in I.$$

Then, by Lemma 2.8,

$$\int_0^1 |\tilde{u}_k'(s) - \tilde{u}'(s)|\, ds \geq C > 0,$$

where C depends only on $R - R_0$ and R'. Thus, we have

$$C \leq \int_0^1 |\tilde{u}_k'(s) - \tilde{u}'(s)|\, ds \leq \varepsilon + \int_0^1 |u_k'(s) - u'(s)|\, ds$$

$$\leq \varepsilon + \|u_k - u\|_{1,p}.$$

Hence $\|u_k - u\|_{1,p} \geq C - \varepsilon$ for all k. Choosing $\varepsilon < C$ this yields $\|u_k - u\|_{1,p} \not\to 0$ as $k \to \infty$, a contradiction.

Thus (2.26) holds and Theorem 2.3 can be applied. $\square$

Proof of Lemma 2.8. We have

$$\chi(t) = \chi(0) + \int_0^t \chi'(\tau)\, d\tau, \quad t \in I.$$

It follows that

$$\|\chi(t)\| - |\chi(s)\|| \leq |\chi(t) - \chi(s)| \leq \left| \int_s^t |\chi'(\tau)|\, d\tau \right| \qquad (2.27)$$

$$\leq M |t - s| \quad \text{for all } t, s \in I.$$

Two cases are possible:

1) for all $t \in I$, $|\chi(t)| \geq a/2$. Then, clearly,

$$\int_0^1 |\chi(s)|\, ds \geq a/2.$$

2) There are $t_1, t_2 \in I$ with $|\chi(t_1)| = a/2$, $|\chi(t_2)| = a$ and $|\chi(t)| \in [a/2, a]$ for all t between t_1 and t_2. Suppose $t_1 < t_2$. Then, if we choose $t = t_1$ and $s = t_2$ in (2.27), we get $t_2 - t_1 \geq a/(2M)$. Also, from (2.27),

$$|\chi(t)| \geq |\chi(t_2)| - M(t_2 - t) = a - M(t_2 - t), \quad t \in [t_1, t_2].$$

Integration from $t_2 - a/(2M)$ to t_2 yields

$$\int_0^1 |\chi(s)|\, ds \geq \int_{t_2 - a/(2M)}^{t_2} |\chi(s)|\, ds \geq 3a^2/(8M).$$

$\square$

Remark 2.4 *In particular, if d is the metric on C_B^1 induced by $\|\cdot\|_{1,p}$ and in addition there is an $r \in (0, R)$ such that in (H3), $\|u\|_{1,\infty} < R-r$ for any solution of $u = (1 - \lambda)\,u_0 + \lambda T\,(u)$, $\lambda \in [0,1]$, then the unique fixed point of T can be approximated by means of the iterative procedure described after Theorem 2.3, where we may use this r and the first approximation $x_0 = u_0$.*

Remark 2.5 *For $p = \infty$, d and d' are equivalent metrics on C_B^1 and Theorem 2.7 is a direct consequence of Theorem 2.4.*

Let $\overline{B}_R = \{u \in E; \ |u| \leq R\}$ and suppose $f : I \times \overline{B}_R^2 \to E$. If f is continuous (resp., L^p-Carathéodory), then the operator

$$F\,(u)\,(t) \; = \; f\,(t, u\,(t), u'\,(t))\,, \quad t \in I$$

is well defined from D_R into $C\,(I; E)$ (resp., $L^p\,(I; E)$) and a function $u \in D_R$ is a classical (resp., Carathéodory) solution of (2.21) if and only if $u = T\,(u)$, where

$$T\,(u) \; = \; u_0 + L^{-1} F\,(u)\,.$$

Recall that u_0 is the unique solution of (2.22).

In order to state an existence and uniqueness principle for (2.21), we embed this problem into a one-parameter family of problems

$$\begin{cases} u'' = \lambda f\,(t, u, u')\,, & t \in I \\ V_1\,(u) = b_1, & V_2\,(u) = b_2, \end{cases} \tag{2.28}$$

where $\lambda \in [0, 1]$.

Theorem 2.9 *Let $R > 0$ and $f : I \times \overline{B}_R^2 \to E$. Assume that $\|u_0\|_{1,\infty} < R$ and the following conditions are satisfied:*

(h1) *f is continuous (resp., $f\,(., u, v)$ is measurable for all $(u, v) \in \overline{B}_R^2$ and $f\,(., 0, 0) \in L^\infty\,(I; E)$);*

(h2) *there exist numbers A_0, $A_1 \geq 0$, function $\phi \in L^\infty\,(I; I)$ and $p \in (1, \infty]$ such that*

$$|f\,(t, u, v) - f\,(t, \bar{u}, \bar{v})| \; \leq \; \phi\,(t)\,(A_0\,|u - \bar{u}| + A_1\,|v - \bar{v}|) \tag{2.29}$$

for a.e. $t \in I$ and all $u, \bar{u}, v, \bar{v} \in \overline{B}_R$, and

$$\rho_p < 1, \qquad (2.30)$$

where

$$\rho_p = A_0 \left[\int_0^1 \left(\int_0^1 |G(t,s)|^q \phi(s)^q \, ds \right)^{p/q} dt \right]^{1/p} \qquad (2.31)$$

$$+ A_1 \left[\int_0^1 \left(\int_0^1 |G_t(t,s)|^q \phi(s)^q \, ds \right)^{p/q} dt \right]^{1/p}$$

for $p < \infty$ $(1/p + 1/q = 1)$, and

$$\rho_\infty = A_0 \max_{t \in I} \int_0^1 |G(t,s)| \phi(s) \, ds + A_1 \max_{t \in I} \int_0^1 |G_t(t,s)| \phi(s) \, ds.$$

(h3) *if $u \in D_R$ solves (2.28) for some $\lambda \in [0,1]$, then $\|u\|_{1,\infty} < R$.*

Then (2.21) has a unique classical (resp., Carathéodory) solution in D_R.

Proof. We shall apply Theorem 2.7. We first note that from (2.29) and $f(.,0,0) \in L^\infty(I;E)$, it follows that f is L^∞-Carathéodory. Now we immediately see that the operator $T(u) = u_0 + L^{-1}F(u)$ is well defined from D_R into $W_B^{2,p}$, $T(D_R)$ is bounded with respect to $\|.\|_{1,\infty}$ and that there is $R' > 0$ such that $|u''(t)| \leq R'$ a.e. on I, for any $u \in T(D_R)$. Hence condition (H1) is satisfied.

Without loss of generality, we may assume that $A_0 > 0$ and $A_1 > 0$. Otherwise, we take $A_0 + \varepsilon$ and $A_1 + \varepsilon$ instead of A_0, A_1 with $\varepsilon > 0$ small enough so that inequality (2.30) remains true. Then, we define a modified L^p-norm on $C^1(I;E)$ by

$$\|u\| = A_0 \|u\|_p + A_1 \|u'\|_p.$$

Clearly, the norms $\|.\|$ and $\|.\|_{1,p}$ are equivalent. Let d be the metric induced by $\|.\|$ on C_B^1. We now check (2.23) and (2.24). Let $u, v \in D_R$. Then using (2.29), we obtain

$$|T(u)(t) - T(v)(t)|$$

$$\leq \int_0^1 |G(t,s)| \, |f(s,u(s),u'(s)) - f(s,v(s),v'(s))| \, ds$$

$$\leq \int_0^1 |G(t,s)| \, \phi(s) \, (A_0 \, |u(s) - v(s)| + A_1 \, |u'(s) - v'(s)|) \, ds$$

$$\leq \left(\int_0^1 |G(t,s)|^q \, \phi(s)^q \, ds \right)^{1/q} \|u - v\|.$$

Also

$$\left| T(u)'(t) - T(v)'(t) \right|$$

$$\leq \int_0^1 |G_t(t,s)| \, |f(s,u(s),u'(s)) - f(s,v(s),v'(s))| \, ds$$

$$\leq \left(\int_0^1 |G_t(t,s)|^q \, \phi(s)^q \, ds \right)^{1/q} \|u - v\|.$$

These clearly yield (2.23) and (2.24), with $\rho = \rho_p$. Hence (H2) is satisfied. Finally (H3) follows from (h3) since a function $u \in D_R$ solves (2.28) if and only if $u = (1 - \lambda) u_0 + \lambda T(u)$. Thus, Theorem 2.7 can be applied. $\square$

Notice that for $p = \infty$ and $\phi = 1$, the result in Theorem 2.9 follows from [[82], Theorem 3.6].

Remark 2.6 *We will compare the contraction condition $\rho_p < 1$ for $p = \infty$ and $p = 2$. Suppose $V_1(u) = u(0)$ and $V_2(u) = u(1)$ and $\phi = 1$. In this case, the Green's function is*

$$G(t,s) = -\begin{cases} t(1-s), & 0 \leq t \leq s \leq 1 \\ s(1-t), & 0 \leq s \leq t \leq 1 \end{cases} \tag{2.32}$$

and routine calculations give

$$\max_{t \in I} \int_0^1 |G(t,s)| \, ds = 1/8, \quad \max_{t \in I} \int_0^1 |G_t(t,s)| \, ds = 1/2,$$

$$\int_0^1 \left(\int_0^1 G(t,s)^2 \, ds \right) dt = 1/90, \quad \int_0^1 \left(\int_0^1 G_t(t,s)^2 \, ds \right) dt = 1/6.$$

It follows that

$$\rho_\infty = A_0/8 + A_1/2 \quad and \quad \rho_2 = A_0/(3\sqrt{10}) + A_1/\sqrt{6}.$$

Thus the contraction condition $\rho_2 < 1$ is less restrictive than $\rho_\infty < 1$.

Remark 2.7 *Other modified L^p-norms on $C^1(I;E)$ are possible and are expected to relax the contraction condition (2.30). For example, we may take the norm*

$$\|u\| = A_0 \|\psi u\|_p + A_1 \|\psi u'\|_p ,$$

where $\psi \in C(I;(0,\infty))$. In this case, the contraction condition becomes

$$A_0 \left[\int_0^1 \psi(t)^p \left(\int_0^1 |G(t,s)|^q \phi(s)^q \psi(s)^{-q} ds \right)^{p/q} dt \right]^{1/p}$$

$$+A_1 \left[\int_0^1 \psi(t)^p \left(\int_0^1 |G_t(t,s)|^q \phi(s)^q \psi(s)^{-q} ds \right)^{p/q} dt \right]^{1/p} < 1$$

for $p < \infty$, and

$$A_0 \max_{t \in I} \int_0^1 \psi(t) |G(t,s)| \phi(s) \psi(s)^{-1} ds$$

$$+A_1 \max_{t \in I} \int_0^1 \psi(t) |G_t(t,s)| \phi(s) \psi(s)^{-1} ds < 1$$

for $p = \infty$.

For such tricks of contraction, we refer the interested reader to [62].

We now consider the case when $\mathcal{B}$ means: $u(0) = u(1) = 0$ and $C_\mathcal{B}^1$ is simply denoted by C_0^1. Then we may choose as d, the metric induced by the norm $\|u\| = \|u'\|_p$. This norm is equivalent to $\|.\|_{1,p}$ since by the Wirtinger-Poincaré inequality

$$\|u\|_p \le \sigma_p \|u'\|_p , \quad u \in C_0^1, \tag{2.33}$$

where

$$1/\sigma_p = \inf \left\{ \|u'\|_p / \|u\|_p ; \ u \in C_0^1, \ u \not\equiv 0 \right\}$$

(see [17] for example). The corresponding existence principle is then the following result.

Theorem 2.10 *Suppose that $\mathcal{B}$ means $u(0) = u(1) = 0$ and all the assumptions of Theorem 2.9 are satisfied with*

$$\rho_p = (A_0 \sigma_p + A_1) \left\{ \int_0^1 \left[\int_0^1 |G_t(t,\tau)|^q \phi(\tau)^q d\tau \right]^{p/q} dt \right\}^{1/p}$$

for $p \in (1, \infty)$ and

$$\rho_\infty = (A_0 \sigma_\infty + A_1) \max_{t \in I} \int_0^1 |G_t(t, \tau)| \phi(\tau) \, d\tau$$

for $p = \infty$. Then (2.21) has a unique classical (resp., Carathéodory) solution in D_R.

In particular, if $p = 2$ and $\phi = 1$, we can prove a more exact result which is based on the Wirtinger and Opial inequalities.

Proposition 2.11 *Let $(E, |.|)$ be a Banach space. If $u \in W_0^{1,2}(I; E)$, then*

$$\int_0^1 |u(t)|^2 \, dt \le \frac{1}{\pi^2} \int_0^1 |u'(t)|^2 \, dt \qquad (2.34)$$

(Wirtinger's inequality) and

$$\int_0^1 |u(t)| \, |u'(t)| \, dt \le \frac{1}{4} \int_0^1 |u'(t)|^2 \, dt \qquad (2.35)$$

(Opial's inequality).

Proof. **1)** The eigenvalue problem

$$\begin{cases} u'' + \lambda u = 0, & t \in I \\ u(0) = u(1) = 0 \end{cases}$$

has solutions $\lambda_k = (k\pi)^2$, $u_k(t) = \sqrt{2} \sin k\pi t$, $k = 1, 2, \dots$. It is known that the sequences $(u_k)_{k \ge 1}$, $\left(\lambda_k^{-1/2} u_k\right)_{k \ge 1}$ are orthonormal and complete in $\left(L^2(I; \mathbf{R}), \langle ., . \rangle_2\right)$ and $\left(W_0^{1,2}(I; \mathbf{R}), \langle ., . \rangle_{0,2}\right)$. Recall that $\langle u, v \rangle_{0,2} = \langle u', v' \rangle_2$. Consequently, if $u \in L^2(I; E)$, then $|u| \in L^2(I; \mathbf{R})$ and

$$|u| = \sum_{k=1}^\infty \langle |u|, u_k \rangle_2 \, u_k.$$

Also, if $u \in W_0^{1,2}(I; E)$, then $|u|$ is absolutely continuous like u, and a direct computation shows that $\left| |u|' \right| \le |u'|$ a.e. on I. It follows that $|u| \in W_0^{1,2}(I; \mathbf{R})$ and consequently,

$$|u| = \sum_{k=1}^\infty \lambda_k^{-1} \langle |u|, u_k \rangle_{0,2} \, u_k.$$

Now Parseval's equality implies that

$$\int_0^1 |u'|^2 \, dt \geq \int_0^1 \left(|u|'\right)^2 dt = \sum_{k=1}^{\infty} \lambda_k^{-1} \langle |u|, u_k \rangle_{0,2}^2$$

$$= \sum_{k=1}^{\infty} \lambda_k \langle |u|, u_k \rangle_2^2 \geq \pi^2 \sum_{k=1}^{\infty} \langle |u|, u_k \rangle_2^2 = \pi^2 \int_0^1 |u|^2 \, dt.$$

1') An elementary proof of Wirtinger's inequality is possible if E is a Hilbert space and $u \in C^1(I; E)$ with $u(0) = u(1) = 0$. In this case we derive (2.34) from the following integral inequalities

$$\int_a^b |u(t) - u(a)|^2 \, dt \leq 4 \left(\frac{b-a}{\pi}\right)^2 \int_a^b |u'(t)|^2 \, dt, \qquad (2.36)$$

$$\int_a^b |u(t) - u(b)|^2 \, dt \leq 4 \left(\frac{b-a}{\pi}\right)^2 \int_a^b |u'(t)|^2 \, dt, \qquad (2.37)$$

which are true for any $u \in C^1([a,b]; E)$. To prove (2.36), we first compute the derivative of

$$D(t) = 2 \frac{b-a}{\pi} |u(t) - u(a)|^2 \cot \frac{\pi(t-a)}{2(b-a)}.$$

We have

$$D'(t) = 4 \frac{b-a}{\pi} \langle u'(t), u(t) - u(a) \rangle \cot \frac{\pi(t-a)}{2(b-a)}$$
$$- |u(t) - u(a)|^2 \left(1 + \cot^2 \frac{\pi(t-a)}{2(b-a)}\right).$$

This can be rewritten as

$$D'(t) = 4 \left(\frac{b-a}{\pi}\right)^2 |u'(t)|^2 - |u(t) - u(a)|^2$$
$$- \left|2 \frac{b-a}{\pi} u'(t) - (u(t) - u(a)) \cot \frac{\pi(x-a)}{2(b-a)}\right|^2.$$

Hence

$$D'(t) \leq 4 \left(\frac{b-a}{\pi}\right)^2 |u'(t)|^2 - |u(t) - u(a)|^2.$$

Integration yields

$$D\left(b\right) - D\left(a+0\right) \leq 4\left(\frac{b-a}{\pi}\right)^2 \int_a^b \left|u'\left(t\right)\right|^2 dt - \int_a^b \left|u\left(t\right) - u\left(a\right)\right|^2 dt.$$

This yields (2.36) since $D\left(b\right) = D\left(a+0\right) = 0$. Inequality (2.37) is proved similarly.

To obtain (2.34), apply (2.36) on $[0, 1/2]$, (2.37) on $[1/2, 1]$ and then piece the results together.

2) Let

$$v\left(t\right) = \int_0^t \left|u'\left(s\right)\right| ds \quad \text{and} \quad w\left(t\right) = \int_t^1 \left|u'\left(s\right)\right| ds. \tag{2.38}$$

Clearly

$$v'\left(t\right) = -w'\left(t\right) = \left|u'\left(t\right)\right|, \quad \text{for a.e. } t \in I.$$

We claim that

$$\left|u\left(t\right)\right| \leq v\left(t\right), \quad \left|u\left(t\right)\right| \leq w\left(t\right), \quad \text{for all } t \in I. \tag{2.39}$$

Indeed, $h\left(t\right) = v\left(t\right) - \left|u\left(t\right)\right|$ is absolutely continuous on I, $h\left(0\right) = 0$ and since $\left|\left|u\right|'\right| \leq \left|u'\right|$ a.e. on I, we have $h'\left(t\right) \geq 0$ for a.e. $t \in I$. Then, from

$$h\left(t\right) = \int_0^t h'\left(s\right) ds,$$

it follows that $h\left(t\right) \geq 0$ for all $t \in I$. Hence $\left|u\left(t\right)\right| \leq v\left(t\right)$ on I. The second inequality can be proved similarly.

From (2.38) and (2.39) we deduce that

$$\int_0^{1/2} \left|u\left(t\right)\right| \left|u'\left(t\right)\right| dt \leq \int_0^{1/2} v\left(t\right) v'\left(t\right) dt = \frac{1}{2} v\left(1/2\right)^2,$$

$$\int_{1/2}^1 \left|u\left(t\right)\right| \left|u'\left(t\right)\right| dt \leq -\int_{1/2}^1 w\left(t\right) w'\left(t\right) dt = \frac{1}{2} w\left(1/2\right)^2.$$

Thus we have the inequality

$$\int_0^1 \left|u\left(t\right)\right| \left|u'\left(t\right)\right| dt \leq \frac{1}{2}\left(v\left(1/2\right)^2 + w\left(1/2\right)^2\right). \tag{2.40}$$

On the other hand, from Hölder's inequality, we obtain

$$v\left(1/2\right)^2 = \left(\int_0^{1/2} \left|u'\left(t\right)\right| dt\right)^2 \leq \frac{1}{2}\int_0^{1/2} \left|u'\left(t\right)\right|^2 dt, \tag{2.41}$$

$$w\left(1/2\right)^2 = \left(\int_{1/2}^1 |u'\left(t\right)|\, dt\right)^2 \leq \frac{1}{2}\int_{1/2}^1 |u'\left(t\right)|^2\, dt. \qquad (2.42)$$

Inequalities (2.40)-(2.42) yield (2.35). □

Now we state and prove the following existence and uniqueness principle for (2.21).

Theorem 2.12 *Let* E *be a Banach space and suppose that* $\mathcal{B}$ *means* $u\left(0\right) = u\left(1\right) = 0$. *Let* $f : I \times \overline{B}_R^2 \to E$ *and assume that the following conditions are satisfied:*

(a1) f *is continuous (resp.,* $f\left(.,u,v\right)$ *is measurable for all* $\left(u,v\right) \in \overline{B}_R^2$ *and* $f\left(.,0,0\right) \in L^\infty\left(I;E\right)$*);*

(a2) *there exist numbers* A_0, $A_1 \geq 0$ *such that*

$$|f\left(t,u,v\right) - f\left(t,\bar{u},\bar{v}\right)| \leq A_0\,|u - \bar{u}| + A_1\,|v - \bar{v}| \qquad (2.43)$$

for a.e. $t \in I$ *and all* $u, \bar{u}, v, \bar{v} \in \overline{B}_R$, *and*

$$\frac{A_0^2}{\pi^4} + \frac{A_1^2}{\pi^2} + \frac{A_0 A_1}{2\pi^2} < 1; \qquad (2.44)$$

(a3) *if* $u \in D_R$ *solves (2.28) for some* $\lambda \in [0,1]$, *then* $\|u\|_{1,\infty} < R$.

Then (2.21) has a unique classical (resp., Carathéodory) solution in D_R.

Proof. We apply Theorem 2.7, where d is the metric on C_0^1 induced by the norm $\|u\| = \|u'\|_2$ and $p = 2$. To show (2.24), we use the Wirtinger's inequality, Opial's inequality and the following inequality

$$\left\|L^{-1}v\right\|_2 \leq \frac{1}{\pi^2}\,\|v\|_2, \quad v \in L^2\left(I;E\right). \qquad (2.45)$$

This can be proved as follows. Let $u = L^{-1}v$. Then $u'' = v$, and if we multiply by u and integrate on I, we get

$$\|u'\|_2^2 = -\langle v, u\rangle_2 \leq \|v\|_2\,\|u\|_2.$$

Using Wirtinger's inequality, this yields

$$\|u'\|_2^2 \leq \frac{1}{\pi}\,\|v\|_2\,\|u'\|_2,$$

and so $\|u'\|_2 \leq (1/\pi) \|v\|_2$. This together with (2.34) yields (2.45). Then, we obtain

$$d\left(T\left(u\right), T\left(v\right)\right) = \left\|\left(L^{-1}\left(F\left(u\right) - F\left(v\right)\right)\right)'\right\|_2$$

$$= \left\{\left(F\left(v\right) - F\left(u\right), L^{-1}\left(F\left(u\right) - F\left(v\right)\right)\right)_2\right\}^{1/2} \leq \frac{1}{\pi} \left\|F\left(u\right) - F\left(v\right)\right\|_2$$

$$\leq \frac{1}{\pi}\left[\int_0^1 \left(A_0 \left|u - v\right| + A_1 \left|u' - v'\right|\right)^2 dt\right]^{1/2}$$

$$\leq \frac{1}{\pi}\left[A_0^2 \left\|u - v\right\|_2^2 + A_1^2 \left\|u' - v'\right\|_2^2 + \frac{A_0 A_1}{2}\left\|u' - v'\right\|_2^2\right]^{1/2}$$

$$\leq \left(\frac{A_0^2}{\pi^4} + \frac{A_1^2}{\pi^2} + \frac{A_0 A_1}{2\pi^2}\right)^{1/2} \left\|u' - v'\right\|_2$$

$$= \left(\frac{A_0^2}{\pi^4} + \frac{A_1^2}{\pi^2} + \frac{A_0 A_1}{2\pi^2}\right)^{1/2} d\left(u, v\right).$$

$\square$

We mention that condition (2.44) was obtained by Hai and Schmitt [62] and used when f is defined on the entire set $I \times E^2$. Therefore, our technique based on the use of two metrics makes it possible to show that certain results involving conditions derived when working with energy L^p-norms extend to the situation when f is defined, or has the required properties, only on a bounded region.

Remark 2.8 *As we have already remarked, for $p = \infty$, Theorem 2.7 is a consequence of Theorem 2.4. For an arbitrary $p < \infty$, if we look at the second proof of Proposition 2.2, we might think of using Theorem 2.4 where we now work in the completion of C_B^1 with respect to d. For example, when B means $u\left(0\right) = u\left(1\right) = 0$, the completion of C_B^1 is the Sobolev space $W_0^{1,p}\left(I; E\right)$. It is easily seen that such an approach has a major drawback, namely the bounded domain of T.*

Remark 2.9 *In the case that f is independent of u' and V_j, $j = 1, 2$, are linear continuous from $C\left(I; E\right)$ into E, we can regard T as a mapping from $D_R^0 = \{u \in C_B; \|u\|_\infty \leq R\} \subset C_B$ into C_B, where*

$$C_B = \{u \in C\left(I; E\right); V_j\left(u\right) = b_j, j = 1, 2\}.$$

It is easy to obtain variants of Theorems 2.7 and 2.9 in which all refer-ence to u' is dropped and the norms $\|\cdot\|_\infty$, $\|\cdot\|_p$ are used instead of $\|\cdot\|_{1,\infty}$ and $\|\cdot\|_{1,p}$ respectively.

Example. Consider the boundary value problem

$$\begin{cases} u'' = f(u), & t \in I \\ u(0) = u(1) = 0. \end{cases} \qquad (2.46)$$

Assume that for some $R > 0$, $f \in C(\overline{B}_R; E)$,

$$\sup\{|f(u)|;\ |u| \le R\} \le 8R$$

and there exists $A_0 < 3\sqrt{10}$ with

$$|f(u) - f(v)| \le A_0 |u - v| \quad \text{for all } u, v \in \overline{B}_R.$$

Then (2.46) has a unique solution (with max-norm at most R). If in addition,

$$\sup\{|f(u)|;\ |u| \le R\} < 8(R - r)$$

for some $0 < r < R$, then the unique solution can be approximated by the iterative procedure described after Theorem 2.3, where: $\rho = A_0/(3\sqrt{10})$, $x_0 \equiv 0$, $H(.,\lambda) = \lambda T$ and d is the metric on $C_0 = \{u \in C(I; E);\ u(0) = u(1) = 0\}$ induced by $\|\cdot\|_2$. According to Remark 2.2, since $\|\cdot\|_2 \le \|\cdot\|_\infty$, it suffices that the estimates in the iterative procedure be made with respect to the metric d' induced by $\|\cdot\|_\infty$.

The above example shows that the continuation principle for contrac-tions applies to problems with superlinear nonlinearity provided that a Lipschitz condition holds in some bounded set. In particular, problem (2.46) for $E = \mathbf{R}$ and $f(u) = -e^u$ was discussed in [82].

Finally, we present a sufficient condition so that (h3) in Theorem 2.9 holds. We discuss the case when E is a Hilbert space and $\mathcal{B}$ are the Sturm-Liouville boundary conditions

$$\mathcal{B} : \begin{cases} u(0) - au'(0) = 0 \\ u(1) + bu'(1) = 0, \end{cases} \qquad (2.47)$$

where $a, b \ge 0$ with $a + b > 0$. The condition is expressed in terms of an inequality of Hartman type (see [63] and [153]).

Theorem 2.13 *Let* $(E, \langle ., . \rangle)$ *be a Hilbert space,* $R > 0$ *and* $f :$ *$I \times \overline{B}_R^2 \to E$. Assume that the following conditions are satisfied:*

(i) f *is continuous;*

(ii) *there exist numbers* $A_0 \geq 0$, $A_1 > 0$, *function* $\phi \in C(I; I)$ *and* $p \in (1, \infty]$ *such that*

$$|f(t, u, v) - f(t, \bar{u}, \bar{v})| \leq \phi(t)(A_0|u - \bar{u}| + A_1|v - \bar{v}|)$$

for all $t \in I$, *all* $u, \bar{u}, v, \bar{v} \in \overline{B}_R$, *and*

$$\rho_p < 1,$$

where

$$\rho_p = A_0 \left[\int_0^1 \left(\int_0^1 |G(t,s)|^q \phi(s)^q \, ds \right)^{p/q} dt \right]^{1/p}$$

$$+ A_1 \left[\int_0^1 \left(\int_0^1 |G_t(t,s)|^q \phi(s)^q \, ds \right)^{p/q} dt \right]^{1/p}$$

for $p < \infty$ $(1/p + 1/q = 1)$, *and*

$$\rho_\infty = A_0 \max_{t \in I} \int_0^1 |G(t,s)| \phi(s) \, ds + A_1 \max_{t \in I} \int_0^1 |G_t(t,s)| \phi(s) \, ds;$$

(iii) *there exists* $r \in (0, R]$ *with*

$$\langle u, f(t, u, v) \rangle + |v|^2 > 0, \quad \text{for } t \in I \tag{2.48}$$

whenever $r \leq |u| \leq R$, $|v| \leq R$ *and* $\langle u, v \rangle = 0$;

(iv) $e^{A_1}(A_0 r + A_1 r_0 + m) \leq A_0 r + A_1 R + m$, *where* $m = \|f(., 0, 0)\|_\infty$ *and* $r_0 = r/\max\{a, b\}$.

Then (2.21)-(2.47) has a unique classical solution satisfying $\|u\|_{1,\infty} < R$.

Proof. The conclusion follows from Theorem 2.9 if we show that condition (h3) is satisfied. For this, let u be any solution to (2.28) for some $\lambda \in (0,1]$. First we show that $\|u\|_\infty < R$. Let $w(t) = |u(t)|^2 / 2$. Then $w'(t) = \langle u(t), u'(t) \rangle$ and

$$w''(t) = \lambda \langle u(t), f(t, u(t), u'(t)) \rangle + |u'(t)|^2, \quad \text{a.e. } t \in I.$$

Let $t_0 \in I$ with $\|u\|_\infty = |u(t_0)|$. If $0 < t_0 < 1$, then $w'(t_0) = 0$ and $w''(t_0) \le 0$ and taking into account (2.48), we find that $|u(t_0)| < r$. Now suppose $t_0 = 0$. Since $w(0)$ is the maximum of w, we have $w'(0) = \langle u(0), u'(0) \rangle \le 0$ and using the first boundary condition in (2.47), we obtain

$$0 \ge a \langle u(0), u'(0) \rangle = |u(0)|^2.$$

Hence $u(0) = 0$ and as a result, inequality $|u(t_0)| < r$ holds. Similarly, the same conclusion holds if $t_0 = 1$. Therefore, $\|u\|_\infty < r \le R$.

Next we show that $\|u'\|_\infty < R$. Let us first remark that $|u'(t_0)| < r_0$ for $t_0 = 0$ or $t_0 = 1$. Indeed, if $a = \max\{a, b\} > 0$, then

$$|u'(0)| = |u(0)|/a < r/a = r_0.$$

Similarly, if $b = \max\{a, b\} > 0$, then

$$|u'(1)| = |u(1)|/b < r/b = r_0.$$

Further, suppose that $\|u'\|_\infty = R$. Let $v(t) = |u'(t)|$ and take $t_1 \in I$ with $v(t_1) = R$. Then, we have

$$v(t)|v'(t)| = |\langle u'(t), u''(t) \rangle| = \lambda |\langle u'(t), f(t, u(t), u'(t)) \rangle|$$

$$\le |u'(t)|(A_0 |u(t)| + A_1 |u'(t)| + |f(t, 0, 0)|)$$

$$\le v(t)(A_0 r + A_1 v(t) + m).$$

Assume $t_0 = 0$. Then, the inequality $v'(t) \le A_0 r + A_1 v(t) + m$ gives

$$\int_{t_0}^{t_1} \frac{v'(s)}{A_0 r + A_1 v(s) + m} \, ds \le t_1 - t_0 \le 1.$$

But, since $v(t_0) < r_0$ and $v(t_1) = R$,

$$\int_{t_0}^{t_1} \frac{v'(s)}{A_0 r + A_1 v(s) + m} \, ds > \int_{r_0}^{R} \frac{dt}{A_0 r + A_1 t + m}$$

$$= \frac{1}{A_1} \ln \frac{A_0 r + A_1 R + m}{A_0 r + A_1 r_0 + m}.$$

These yield

$$\frac{1}{A_1} \ln \frac{A_0 r + A_1 R + m}{A_0 r + A_1 r_0 + m} < 1,$$

which contradicts (iv). The same contradiction is derived when $t_0 = 1$, if the inequality $v'(t) \leq A_0 r + A_1 v(t) + m$ is replaced by $-v'(t) \leq A_0 r + A_1 v(t) + m$. Thus, $\|u'\|_\infty < R$ and so $\|u\|_{1,\infty} < R$. $\square$

For other results of this nature we refer the reader to [46] and [82].

3. Continuation Theorems for Nonexpansive Maps

3.1 Continuation Theorems

Let $(X, |\,.\,|)$ be a Banach space. A map $T : D \subset X \to X$ is said to be *nonexpansive* if

$$|T(x) - T(y)| \leq |x - y| \quad \text{for all } x,\, y \in D.$$

The question of interest is does a nonexpansive self-map of a given set D have a fixed point? The answer is yes if D is a nonempty closed bounded and convex set of a uniformly convex Banach space (see [18], [54] and [71]).

A Banach space $(X, |\,.\,|)$ is said to be *uniformly convex* if for each $\varepsilon > 0$ there exists $\delta_\varepsilon > 0$ with

$$|x + y| \leq 2\,(1 - \delta_\varepsilon)$$

for all $x, y \in X$ satisfying $|x| \leq 1$, $|y| \leq 1$ and $|x - y| \geq \varepsilon$.

Recall that any uniformly convex Banach space is reflexive (see [[33], Proposition 3.12.1]) and any Hilbert space is uniformly convex since for $|x| \leq 1$ and $|y| \leq 1$, one has $|x + y|^2 + |x - y|^2 = 2(|x|^2 + |y|^2) \leq 4$.

Let us first recall the Browder-Göhde-Kirk fixed point theorem.

Proposition 3.1 *Let* $(X, |\,.\,|)$ *be a uniformly convex Banach space,* $D \subset X$ *nonempty closed bounded and convex. Assume* $T : D \to D$ *is nonexpansive. Then* T *has at least one fixed point.*

43

For the proof we recommend [53] or [150].

Now we state and prove the corresponding Leray-Schauder type theorem [104], [139].

Theorem 3.2 *Let* $(X, |\cdot|)$ *be a uniformly convex Banach space,* $U \subset X$ *open bounded convex with* $0 \in U$ *and* $T : \overline{U} \to X$ *nonexpansive. Assume*

$$\lambda T(x) \neq x \quad \text{for all } x \in \partial U \text{ and } \lambda \in [0,1]. \qquad (3.1)$$

Then T *has at least one fixed point in* U.

For the proof we need the following two lemmas due to Browder [19].

Lemma 3.3 *Let* $(X, |\cdot|)$ *be a uniformly convex Banach space,* $D \subset X$ *bounded convex and* $T : D \to X$ *nonexpansive. Then, for each* $\varepsilon > 0$ *there exists* $\delta = \delta(\varepsilon) > 0$ *such that* $|x - T(x)| \leq \varepsilon$ *for all* $x = (1 - \lambda) x_0 + \lambda x_1$, *where* $\lambda \in (0,1)$, $x_0, x_1 \in D$, $|x_0 - T(x_0)| \leq \delta$ *and* $|x_1 - T(x_1)| \leq \delta$.

Proof. If $|x_0 - x_1| < \varepsilon/3$, then for $x = x_\lambda := (1 - \lambda) x_0 + \lambda x_1$, we have

$$|x - T(x)| \leq |x - x_0| + |x_0 - T(x_0)| + |T(x_0) - T(x)|$$

$$\leq |x_0 - T(x_0)| + 2|x - x_0| \leq \delta + 2\varepsilon/3 \leq \varepsilon$$

provided that $\delta \leq \varepsilon/3$. Thus, we have only to deal with the case $|x_0 - x_1| \geq \varepsilon/3$. If $\lambda < \varepsilon/(3d)$, where $d = \operatorname{diam} D$, then $|x_\lambda - x_0| = \lambda |x_0 - x_1| < \varepsilon/3$. Then, as above, we obtain $|x_\lambda - T(x_\lambda)| \leq \varepsilon$. Furthermore, we assume $\lambda \geq \varepsilon/(3d)$. If $1 - \lambda < \varepsilon/(3d)$, then by a similar argument, we get $|x_\lambda - T(x_\lambda)| \leq \varepsilon$ where $\delta \leq \varepsilon/3$. Now we examine the case: $\varepsilon/(3d) \leq \lambda \leq 1 - \varepsilon/(3d)$ and $|x_0 - x_1| \geq \varepsilon/3$. We have

$$|x_0 - T(x_\lambda)| \leq |x_0 - T(x_0)| + |T(x_0) - T(x_\lambda)|$$

$$\leq \delta + \lambda |x_0 - x_1|$$

and similarly

$$|x_1 - T(x_\lambda)| \leq \delta + (1 - \lambda) |x_0 - x_1|.$$

Let

$$y_0 = \lambda^{-1} |x_0 - x_1|^{-1} (T(x_\lambda) - x_0)$$

$$y_1 = (1 - \lambda)^{-1} |x_0 - x_1|^{-1} (x_1 - T(x_\lambda)) .$$

Then it is easily seen that

$$|y_0| \leq 1 + 9d\varepsilon^{-2}\delta, \quad |y_1| \leq 1 + 9d\varepsilon^{-2}\delta,$$

$$|\lambda y_0 + (1 - \lambda) y_1| = 1.$$

For δ small enough, by the uniform convexity of X, we deduce that $|y_0 - y_1| \leq \varepsilon/d$. It follows that

$$|x_\lambda - T(x_\lambda)| = |(1 - \lambda)(T(x_\lambda) - x_0) - \lambda(x_1 - T(x_\lambda))|$$

$$= \lambda(1 - \lambda) |x_0 - x_1| |y_0 - y_1| \leq d \cdot \varepsilon/d = \varepsilon.$$

$\square$

Lemma 3.4 Let $(X, |.|)$ be a uniformly convex Banach space, $D \subset X$ closed bounded convex and $T : D \to X$ nonexpansive. If $(x_k) \subset D$, $x_k \to x_0$ weakly and $x_k - T(x_k) \to y_0$ strongly as $k \to \infty$, then $x_0 - T(x_0) = y_0$.

Proof. We may assume without loss of generality that $y_0 = 0$. Let $\varepsilon_k = |x_k - T(x_k)|$. Since $\varepsilon_k \to 0$ as $k \to \infty$, we may suppose that $\varepsilon_k \leq \delta(\varepsilon_{k-1}) \leq \varepsilon_{k-1}$ for any k, where $\delta(\varepsilon)$ is given as in Lemma 3.3. Since $x_0 \in D_k := \overline{\mathrm{conv}} \{x_j; j \geq k\}$ for every k, it follows that in order to prove $x_0 - T(x_0) = 0$, it suffices to show

$$|x - T(x)| \leq \varepsilon_{k-1} \quad \text{for all } x \in D_k. \tag{3.2}$$

Each element of D_k is the strong limit of some sequence of finite convex combinations of elements in $\{x_j; j \geq k\}$. Thus it is sufficient to prove inequality (3.2) only for the elements of $D_k^m := \mathrm{conv} \{x_j; k \leq j \leq m\}$, $m \geq k$. We use induction with respect to k decreasing: For $k = m$, $|x_m - T(x_m)| = \varepsilon_m \leq \varepsilon_{m-1}$ and (3.2) holds. Let us now assume that (3.2) is true for all $x \in D_k^m$. Let $y \in D_{k-1}^m$. Then $y = (1 - \lambda) x_{k-1} + \lambda x$ for some $x \in D_k^m$ and $\lambda \in [0, 1]$. We have

$$|x - T(x)| \leq \varepsilon_{k-1} \leq \delta(\varepsilon_{k-2})$$

and

$$|x_{k-1} - T(x_{k-1})| = \varepsilon_{k-1} \leq \delta(\varepsilon_{k-2}) .$$

From Lemma 3.3, we have $|y - T(y)| \leq \varepsilon_{k-2}$ which proves (3.2) on D_{k-1}^m. $\square$

Proof of Theorem 3.2. From Theorem 2.4, for each $\lambda \in (0,1)$, there exists a unique $x_\lambda \in U$ with $x_\lambda - \lambda T(x_\lambda) = 0$. Denote by x_k the element x_λ for $\lambda = 1 - 1/k$. Taking if necessary a subsequence, we may suppose that $x_k \to x_0$ weakly as $k \to \infty$. On the other hand, since

$$x_k - (1 - 1/k) T(x_k) = 0,$$

it follows that $x_k - T(x_k) \to 0$ strongly as $k \to \infty$. Now Lemma 3.4 guarantees that $x_0 - T(x_0) = 0$. $\square$

If X is a Hilbert space, the convexity of U is not necessary and the result in Theorem 3.2 is due to Gatica and Kirk [50]. In addition, a simpler proof is possible [139], [141].

Theorem 3.5 *Let $(X, \langle ., . \rangle)$ be a Hilbert space, $U \subset X$ open bounded with $0 \in U$ and $T : \overline{U} \to X$ nonexpansive. Assume*

$$\lambda T(x) \neq x \quad \text{for all } x \in \partial U \text{ and } \lambda \in [0,1].$$

Then T has at least one fixed point in U.

Proof. The sequence (x_k) obtained in the proof of Theorem 3.2 satisfies

$$\left\langle (k-1)^{-1} x_k - (m-1)^{-1} x_m, x_k - x_m \right\rangle$$
$$= \langle T(x_k) - T(x_m), x_k - x_m \rangle - |x_k - x_m|^2 \leq 0$$

for $k, m > 1$. Let $a_k = (k-1)^{-1}$. The following identity can easily be checked

$$2 \langle a_k x_k - a_m x_m, x_k - x_m \rangle = (a_k + a_m) |x_k - x_m|^2$$
$$+ (a_k - a_m) \left(|x_k|^2 - |x_m|^2 \right)$$

It follows that

$$0 \leq (a_k + a_m) |x_k - x_m|^2 \leq (a_k - a_m) \left(|x_m|^2 - |x_k|^2 \right).$$

From this, since (a_k) is decreasing, we see that $(|x_k|)$ is increasing. This sequence is also bounded because U is bounded. Hence $(|x_k|)$ is convergent. Now, using

$$|x_k - x_m|^2 \leq \left(|x_m|^2 - |x_k|^2 \right) (a_k - a_m) / (a_k + a_m)$$

we deduce that (x_k) is convergent. Clearly, its limit is a fixed point of T. $\square$

For U an open ball $B_R(0;X)$, Theorem 3.5 was proved in [36] using the radial projection.

3.2 Elements of Geometry of Sobolev Spaces

Here are some basic geometry results for the $L^p(I;E)$ [33] and $W^{m,p}(I;E)$ spaces, where $I = [0,1]$ and E is a Banach space.

Proposition 3.6 *Let E be a uniformly convex Banach space and $1 < p < \infty$. Then the space $L^p(I;E)$ is uniformly convex.*

For the proof we need the following lemma.

Lemma 3.7 *Let $(X, |.|)$ be a uniformly convex Banach space and $1 < p < \infty$. Then, for each $\varepsilon > 0$, there exists $\delta_p = \delta_p(\varepsilon) > 0$ such that*

$$\left| \frac{1}{2}(x+y) \right|^p \leq (1 - \delta_p) \frac{1}{2} (|x|^p + |y|^p) \tag{3.3}$$

for all $x, y \in \overline{B}_1(0;X)$ with $|x - y| \geq \varepsilon$.

Proof. For $1 < p < \infty$, $t = 1$ is the unique point of minimum of the function $\phi(t) = (1 + t^p)/(1 + t)^p$, $t \geq 0$ and, $\phi(1) = 2^{1-p}$.

We first note that it suffices to prove (3.3) for $|x| = 1$, $|y| \leq 1$ and $|x - y| \geq \varepsilon$. Indeed, if $1 \geq |x| \geq |y|$, $|x - y| \geq \varepsilon$ and we set $x' = |x|^{-1} x$, $y' = |x|^{-1} y$, then $|x'| = 1$, $|y'| \leq 1$ and $|x' - y'| \geq |x|^{-1} \varepsilon \geq \varepsilon$. Now inequality (3.3) for x, y follows from the corresponding inequality for x', y', because of the homogeneity.

Suppose there exists $\varepsilon > 0$ such that the conclusion be false. Then there exist two sequences (x_k) and (y_k) with $|x_k| = 1$, $|y_k| \leq 1$, $|x_k - y_k| \geq \varepsilon$ and

$$2 \left| \frac{1}{2}(x_k + y_k) \right|^p / (|x_k|^p + |y_k|^p) \to 1 \quad \text{as } k \to \infty. \tag{3.4}$$

Then, taking if necessary a subsequence, we may assume $|y_k| \to 1$ as $k \to \infty$. Otherwise, $|y_k| \leq r < 1$ for all k, and by the convexity of $|.|$,

$$\left| \frac{1}{2}(x_k + y_k) \right|^p \leq \frac{1}{2^p} (1 + |y_k|)^p.$$

In addition

$$\phi\left(|y_k|\right) \geq \omega := \min_{t \in [0, r]} \phi\left(t\right) > \phi\left(1\right) = 2^{1-p}.$$

Then

$$2 \left| \frac{1}{2}\left(x_k + y_k\right) \right|^p / \left(1 + |y_k|^p\right) \leq \frac{1}{2^{p-1}\omega} < 1.$$

This contradicts (3.4).

Let $z_k = |y_k|^{-1} y_k$. We have

$$|z_k - y_k| = \left| |y_k|^{-1} y_k - y_k \right| \to 0 \quad \text{as} \quad k \to \infty.$$

Hence, for k large enough, $|y_k - z_k| \leq \varepsilon/2$ and consequently,

$$|x_k - z_k| \geq |x_k - y_k| - |y_k - z_k| \geq \varepsilon - \varepsilon/2 = \varepsilon/2.$$

On the other hand, from (3.4), we easily obtain that $\left| 2^{-1}\left(x_k + z_k\right) \right| \to 1$ as $k \to \infty$. This contradicts the uniform convexity of X. $\square$

Proof of Proposition 3.6. Clearly, inequality (3.3) remains true for all $x, y \in X$ with $|x - y| \geq \varepsilon \max\left\{|x|, |y|\right\}$.

Let us fix $\varepsilon > 0$ and let $\eta = \delta_p\left(\varepsilon 4^{-1/p}\right)$. Take $u, v \in L^p\left(I; E\right)$ with $\|u\|_p \leq 1$, $\|v\|_p \leq 1$ and $\|u - v\|_p \geq \varepsilon$. Let

$$M = \left\{t \in I;\ \varepsilon^p\left(|u\left(t\right)|^p + |v\left(t\right)|^p\right) \leq 4\left|u\left(t\right) - v\left(t\right)|^p\right\}.$$

If $t \in M$, then

$$|u\left(t\right) - v\left(t\right)| \geq \varepsilon\, 4^{-1/p} \max\left\{|u\left(t\right)|, |v\left(t\right)|\right\}$$

and so

$$\left| \frac{1}{2}\left(u\left(t\right) + v\left(t\right)\right) \right|^p \leq \left(1 - \eta\right) \frac{1}{2}\left(|u\left(t\right)|^p + |v\left(t\right)|^p\right). \qquad (3.5)$$

Since

$$\int_{I \setminus M} |u\left(t\right) - v\left(t\right)|^p\, dt \leq \frac{\varepsilon^p}{4} \int_{I \setminus M} \left(|u\left(t\right)|^p + |v\left(t\right)|^p\right) dt \leq \varepsilon^p/2$$

and $\|u - v\|_p \geq \varepsilon$, we have

$$\int_M |u\left(t\right) - v\left(t\right)|^p\, dt \geq \varepsilon^p/2.$$

Consequently

$$\max\left\{\int_M |u(t)|^p \, dt, \, \int_M |v(t)|^p \, dt\right\} \geq \varepsilon^p 2^{-p-1}. \qquad (3.6)$$

Now (3.5) and (3.6) yield

$$\int_0^1 \left[\frac{1}{2}\left(|u|^p + |v|^p\right) - \left|\frac{1}{2}(u+v)\right|^p\right] dt$$

$$\geq \int_M \left[\frac{1}{2}\left(|u|^p + |v|^p\right) - \left|\frac{1}{2}(u+v)\right|^p\right] dt$$

$$\geq \int_M \eta \frac{1}{2}\left(|u|^p + |v|^p\right) dt \geq \eta \, \varepsilon^p 2^{-p-2}.$$

This implies

$$\int_0^1 \left|\frac{1}{2}(u+v)\right|^p dt \leq \int_0^1 \frac{1}{2}\left(|u|^p + |v|^p\right) dt - \eta \, \varepsilon^p 2^{-p-2}$$

$$\leq 1 - \eta \, \varepsilon^p 2^{-p-2}.$$

□

Corollary 3.8 *Let $(E, |\cdot|)$ be a uniformly convex Banach space, $m \in \mathbf{N}$, $m \geq 1$ and $1 < p < \infty$. Then the Sobolev space $W^{m,p}(I; E)$ equipped with the norm*

$$\|u\| = \left\{\int_0^1 \left(\sum_{j=0}^m \left|u^{(j)}(t)\right|^p\right) dt\right\}^{1/p}$$

is uniformly convex.

Proof. We imbed $W^{m,p}(I; E)$ into $L^p(I; E^{m+1})$ with the identification

$$u = \left(u, u', u'', \ldots, u^{(m)}\right).$$

Thus the problem reduces to the proof of the uniform convexity of the space E^{m+1} equipped with the norm

$$|v|_{E^{m+1}} = \left(\sum_{j=0}^m |v_j|^p\right)^{1/p}, \quad v = (v_0, v_1, \ldots, v_m).$$

On the other hand, E^{m+1} can be imbedded into $L^p(I; E)$ if we identify any $v = (v_0, v_1, ..., v_m) \in E^{m+1}$ with the function

$$u_v(t) = (m+1)^{1/p} v_j \text{ for } t \in \left[\frac{j}{m+1}, \frac{j+1}{m+1} \right), \ j = 0, 1, ..., m.$$

Notice that

$$\|u_v\|_p = |v|_{E^{m+1}}.$$

Then, since $L^p(I; E)$ is uniformly convex, its normed subspace $(E^{m+1}, |\cdot|_{E^{m+1}})$ is uniformly convex too. $\square$

3.3   Sturm-Liouville Problems in Uniformly Convex Banach Spaces

As an example of an application of the abstract results presented in Section 1, let us consider the Sturm-Liouville problem

$$\begin{cases} u'' = f(t, u, u'), & t \in I \\ u(0) - au'(0) = 0 \\ u(1) + bu'(1) = 0, \end{cases} \tag{3.7}$$

where $a, b \geq 0$, and the corresponding one-parameter family of problems

$$\begin{cases} u'' = \lambda f(t, u, u'), & t \in I \\ u(0) - au'(0) = 0 \\ u(1) + bu'(1) = 0, \end{cases} \tag{3.8}$$

where $\lambda \in [0, 1]$. Let G be the Green's function corresponding to operator u'' and the Sturm-Liouville boundary conditions.

We have the following existence principle for (3.7).

Theorem 3.9 *Let $(E, |\cdot|)$ be a uniformly convex Banach space and $f : I \times E^2 \to E$. Assume that the following conditions are satisfied:*

(a) *f is continuous (resp., $f(., u, v)$ is measurable for all $u, v \in E$);*

(b) *there exists $\phi \in L^\infty(I; \mathbf{R})$ and $p \in (1, \infty)$ such that $f(., 0, 0) \in L^p(I; E)$,*

$$|f(t, u, v) - f(t, \bar{u}, \bar{v})| \leq \phi(t) (|u - \bar{u}| + |v - \bar{v}|) \tag{3.9}$$

for a.e. $t \in I$, all u, $\bar{u}$, v, $\bar{v} \in E$, and $\rho \le 1$, where ρ equals

$$2^{1/q} \left\{ \int_0^1 \left[\left(\int_0^1 |G|^q \, \phi(s)^q \, ds \right)^{p/q} + \left(\int_0^1 |G_t|^q \, \phi(s)^q \, ds \right)^{p/q} \right] dt \right\}^{1/p}$$

$(1/p + 1/q = 1)$;

(c) *there is $R > 0$ such that for any solution u of (3.8), one has*

$$\|u\| = \left(\int_0^1 (|u|^p + |u'|^p) \, dt \right)^{1/p} < R.$$

Then (3.7) has at least one classical (resp., Carathéodory) solution.

Proof. The map

$$T = L^{-1}F : W^{1,p}(I; E) \rightarrow W^{1,p}(I; E)$$

is nonexpansive for $\rho \le 1$ with respect to norm $\|.\|$ on $W^{1,p}(I; E)$. Indeed, for every u, $v \in W^{1,p}(I; E)$, we have

$$|T(u)(t) - T(v)(t)|$$

$$\le \int_0^1 |G(t,s)| \, \phi(s) \, (|u(s) - v(s)| + |u'(s) - v'(s)|) \, ds$$

$$\le \left(\int_0^1 |G|^q \, \phi^q ds \right)^{1/q} \left(\int_0^1 [|u - v| + |u' - v'|]^p \, ds \right)^{1/p}$$

Using Hölder's inequality:

$$(a_1 b_1 + a_2 b_2) \le (a_1^p + a_2^p)^{1/p} (b_1^q + b_2^q)^{1/q},$$

which is true for all nonnegative real numbers a_1, a_2, b_1, b_2, we deduce that

$$|T(u)(t) - T(v)(t)| \le 2^{1/q} \left(\int_0^1 |G|^q \, \phi^q ds \right)^{1/q} \|u - v\|.$$

Similarly

$$\left| T(u)'(t) - T(v)'(t) \right| \le 2^{1/q} \left(\int_0^1 |G_t|^q \, \phi^q ds \right)^{1/q} \|u - v\|.$$

These yield

$$\|T(u) - T(v)\| \le \rho \|u - v\|.$$

Thus we may apply Theorem 3.2. □

As a consequence, we have the following existence result.

Theorem 3.10 *Let* $(E, \langle . , . \rangle)$ *be a Hilbert space and* $f : I \times E^2 \to E$ *continuous. Assume that condition (b) of Theorem 3.9 holds with* $\phi \in C(I; \mathbf{R})$. *In addition assume that* $a + b > 0$ *and there exists* $r > 0$ *such that*

$$\langle u, f(t, u, v) \rangle + |v|^2 > 0 \quad \text{for all } t \in I, \tag{3.10}$$

whenever $|u| \geq r$ *and* $\langle u, v \rangle = 0$. *Then there exists at least one* $u \in C^2(I; E)$ *that solves (3.7).*

Proof. We show that condition (c) in Theorem 3.9 is satisfied. Let u be any solution of (3.8) for some $\lambda \in (0, 1]$. As in the proof of Theorem 2.13, we can show that $\|u\|_\infty < r$ and

$$|u'(t_0)| < r_0 = r / \max\{a, b\},$$

where $t_0 = 0$ or $t_0 = 1$. Next we obtain a bound for $\|u'\|_\infty$. Let $v(t) = |u'(t)|$ and $t_1 \in I$ with $v(t_1) = \|u'\|_\infty$. From (3.9), we have

$$|v(t)\, v'(t)| = |\langle u'(t), u''(t) \rangle| \leq \beta\, v(t)\, (r + v(t) + 1),$$

where $\beta = \max \left\{ \max_{t \in I} \phi(t), \ \max_{t \in I} |f(t, 0, 0)| \right\}$. This implies

$$\int_{v(t_0)}^{v(t_1)} \frac{1}{1 + r + t}\, dt = \left| \int_{t_0}^{t_1} \frac{v'(s)}{1 + r + v(s)}\, ds \right| \leq \beta.$$

Thus, there exists $r_1 > 0$ with $v(t_1) < r_1$. Then $\|u\| < R$, where $R = (r^p + r_1^p)^{1/p}$. Therefore, Theorem 3.9 applies. $\square$

4. Theorems of Leray-Schauder Type for Accretive Maps

4.1 Properties of Accretive Maps

Let $(X, |.|)$ be a Banach space and let $(X^*, |.|)$ denote the dual of X. Let $\mathcal{F} : X \to 2^{X^*}$ be the *duality map* ,

$$\mathcal{F}(x) = \left\{ x^* \in X^*; \; x^*(x) = |x|^2 = |x^*|^2 \right\}$$

and let $\langle . , . \rangle_+$, $\langle . , . \rangle_-$ be the *semi-inner products* on X defined by

$$\langle x, y \rangle_+ = |y| \lim_{t \to 0^+} t^{-1} (|y + tx| - |y|)$$

$$\langle x, y \rangle_- = |y| \lim_{t \to 0^-} t^{-1} (|y + tx| - |y|)$$

or equivalently

$$\langle x, y \rangle_+ = \sup \{ y^*(x) ; \; y^* \in \mathcal{F}(y) \}$$

$$\langle x, y \rangle_- = \inf \{ y^*(x) ; \; y^* \in \mathcal{F}(y) \}.$$

The following properties are simple consequences of the definitions (see [[32], Proposition 13.1]):

1)

$$\langle x + y, z \rangle_\pm \le \langle x, z \rangle_\pm + \langle y, z \rangle_+$$

53

$$\langle x + \alpha y, y \rangle_\pm = \alpha |y|^2 + \langle x, y \rangle_\pm$$

$$\left| \langle x, y \rangle_\pm \right| \le |x| |y| ;$$

2) $\langle x, y \rangle_- \le \langle x, y \rangle_+$ with equality if and only if X^* is strictly convex;

3) If X^* is uniformly convex, then $\langle ., . \rangle_- = \langle ., . \rangle_+$ is uniformly continuous on bounded subsets of $X \times X$;

4) If $u : (a, b) \to X$ is differentiable, then $\phi(t) = |u(t)|$ satisfies

$$\phi(t) D^- \phi(t) \le \langle u'(t), u(t) \rangle_-$$

where $D^- \phi(t) = \limsup_{h \to 0+} h^{-1} (\phi(t) - \phi(t - h))$.

A map $F : D \subset X \to X$, is said to be *accretive* if

$$\langle F(x) - F(y), x - y \rangle_+ \ge 0 \quad \text{for all } x, y \in D,$$

strongly accretive if there exists $c > 0$ such that

$$\langle F(x) - F(y), x - y \rangle_+ \ge c |x - y|^2 \quad \text{for all } x, y \in D$$

and *ψ-accretive* if

$$\langle F(x) - F(y), x - y \rangle_+ \ge \psi(|x - y|) |x - y| \quad \text{for } x, y \in D,$$

where $\psi : \mathbf{R}_+ \to \mathbf{R}_+$ is a continuous function satisfying

$$\psi(0) = 0, \ \psi(r) > 0 \ \text{ for } r > 0 \ \text{ and } \liminf_{r \to \infty} \psi(r) > 0.$$

In particular, if $(X, \langle ., . \rangle)$ is a Hilbert space, the duality map is the identity map of X, the semi-inner products $\langle ., . \rangle_-$ and $\langle ., . \rangle_+$ coincide with the inner product $\langle ., . \rangle$ of X and the accretive and strongly accretive maps are the so called *monotone* and *strongly monotone* maps, respectively.

For example, if $T : D \to X$ is a contraction with Lipschitz constant ρ, then $J - T$ is strongly accretive with $c = 1 - \rho$ (here J stands for the identity map on X). Also, if T is nonexpansive, then $J - T$ is accretive. This is the reason that maps of the form $J - F$ with F accretive are called *pseudocontractive*.

Recall that a map is said to be *demicontinuous* if it is continuous from the strong topology to the weak topology.

4.2   Continuation Principles for Accretive Maps

The Leray-Schauder type theorem for strongly accretive maps is due to Morales [96]. The version of this result for ψ-accretive maps (where in addition ψ is supposed to be strictly increasing and $\psi(r) \to \infty$ as $r \to \infty$) appears in [104]. Preceding results were obtained in [50], [72], [155] and [95].

Theorem 4.1 *Let X be a Banach space, U an open subset of X and $T : \overline{U} \to X$. Suppose that $J - T$ is ψ-accretive and that there is $x_0 \in U$ such that*

$$(1 - \lambda) x_0 + \lambda T(x) \neq x \quad \text{for all } x \in \partial U, \lambda \in [0, 1].$$

Assume that T sends bounded sets into bounded sets if $\psi(r) \to \infty$ as $r \to \infty$, and assume $T(\overline{U})$ is bounded otherwise. In addition suppose that one of the following conditions is satisfied:

(a) *T is continuous;*

(b) *T is demicontinuous and X^* is uniformly convex.*

    Then T has a (unique) fixed point in U.

    For the proof we need the following four results:

Lemma 4.2 *Let X be a Banach space, $D = \overline{B}_r(x_0; X)$ and let $F : D \to X$. Suppose that there exists $c \in \mathbf{R}$ such that*

$$\langle F(x) - F(y), x - y \rangle_- \leq c |x - y|^2 \quad \text{for all } x, y \in D.$$

Also assume that $|F(x)| \leq M$ on D. In addition suppose that (a) or (b) holds. Then the initial value problem

$$u' = F(u), \quad u(0) = x_0$$

has a unique solution on $[0, r/M]$.

Proof. Suppose (b) holds. Consider the approximations

$$u_k(t) = x_0 \quad \text{for } t \le 0,$$

$$u_k(t) = x_0 + \int_0^t F(u_k(s - 1/k)) \, ds \quad \text{for } t > 0.$$

Then

$$|u_k(t) - x_0| \le Mt \le r \quad \text{and} \quad |u_k(t) - u_k(s)| \le M |t - s|$$

for all $t, s \in [0, r/M]$. Also

$$u'_k(t) = F(u_k(t - 1/k)).$$

Let $\phi(t) = |u_k(t) - u_j(t)|$ for some fixed k, j. Then $\phi(0) = 0$ and, by 4),

$$\phi(t) D^- \phi(t) \le \left\langle u'_k(t) - u'_j(t), u_k(t) - u_j(t) \right\rangle_-$$

$$\le \left\langle F(u_k(t - 1/k)) - F(u_j(t - 1/j)), u_k(t - 1/k) - u_j(t - 1/j) \right\rangle_-$$

$$+ 2M^2(1/k + 1/j) \le c[\phi(t) + M(1/k + 1/j)]^2 + 2M^2(1/k + 1/j).$$

Since $(a + b)^2 \le 2(a^2 + b^2)$, this implies

$$D^- \phi^2(t) \le 4|c| \phi^2(t) + c_{kj} \quad \text{with } c_{kj} \to 0 \text{ as } k, j \to \infty.$$

It follows that ϕ^2 is not longer than the solution of the initial value problem

$$\rho' = 4|c| \rho + c_{kj}, \quad \rho(0) = 0.$$

Indeed, given $\varepsilon > 0$, let ρ_ε be the solution of

$$\rho'_\varepsilon = 4|c| \rho_\varepsilon + c_{kj} + \varepsilon, \quad \rho_\varepsilon(0) = \varepsilon.$$

Then $0 = \phi^2(0) < \rho_\varepsilon(0)$ and if there was a smallest $t_0 > 0$ with $\phi^2(t_0) = \rho_\varepsilon(t_0)$, then

$$D^- \phi^2(t_0) = \limsup_{h \to 0^+} h^{-1} \left(\phi^2(t_0) - \phi^2(t_0 - h) \right)$$

$$\ge \lim_{h \to 0^+} h^{-1} \left(\rho_\varepsilon(t_0) - \rho_\varepsilon(t_0 - h) \right) = 4|c| \rho_\varepsilon(t_0) + c_{kj} + \varepsilon$$

$$= 4|c| \phi^2(t_0) + c_{kj} + \varepsilon > D^- \phi^2(t_0),$$

a contradiction. Hence, $\phi^2(t) \le \rho_\varepsilon(t)$ and letting $\varepsilon \to 0$ we obtain $\phi^2(t) \le \rho(t)$ on $[0, r/M]$. Thus, $\phi^2(t) \le \alpha c_{kj}$, where

$$\alpha = (4\,|c|)^{-1} \exp(4\,|c|\,r/M) \quad \text{for } c \neq 0 \text{ and}$$
$$\alpha = r/M \quad \text{for } c = 0.$$

Hence (u_k) is Cauchy. Consequently, $u_k(t) \to u(t)$ uniformly on $[0, r/M]$, for some continuous function u. Passing to limit as $k \to \infty$ and taking into account the uniform continuity of $\langle\,.\,,\,.\,\rangle_+$, we obtain that u is the desired solution.

To prove the uniqueness, suppose that u and v are solutions and let $\phi = |u - v|$. Then

$$\phi(t)\, D^-\phi(t) \le c\phi^2(t), \quad \phi(0) = 0.$$

It follows that $\phi(t) \equiv 0$.

Under condition (a), F can be approximated uniformly by locally Lipschitz maps F_k. Then with F_k instead of F, there is a unique u_k with

$$u_k' = F_k(u_k) = F(u_k) + v_k \quad \text{on} \quad [0, r/M].$$

We have

$$|v_k(t)| \le \sup |F(x) - F_k(x)| \to 0 \quad \text{as} \quad k \to \infty.$$

Now use similar estimates as in the first part to see that (u_k) is Cauchy. Pass to the limit to obtain the desired solution. $\square$

Lemma 4.3 *Let X be a Banach space, D a closed subset of X. Assume that $F : D \to X$ is ψ-accretive and demicontinuous. Then $F(D)$ is closed in X.*

Proof. Suppose $F(x_k) \to y \in X$ as $k \to \infty$, where $x_k \in D$. From

$$\psi(|x_k - x_m|)\,|x_k - x_m| \le \langle F(x_k) - F(x_m),\, x_k - x_m \rangle_+$$
$$\le |F(x_k) - F(x_m)|\,|x_k - x_m|$$

we obtain

$$\psi(|x_k - x_m|) \to 0 \quad \text{as} \quad k,\, m \to \infty.$$

Taking into account the properties of ψ, we deduce that (x_k) is a Cauchy sequence. Thus there is $x_0 \in X$ with $x_k \to x_0$ as $k \to \infty$. Since D is closed, $x_0 \in D$ too. By the demicontinuity of F, we obtain $y = F(x_0)$, that is $y \in F(D)$. $\square$

The next result is due to Deimling [30].

Lemma 4.4 *Let X be a Banach space and U an open subset of X. Assume that $F : \overline{U} \to X$ is ψ-accretive and that (a) or (b) holds. Then $F(U)$ is open in X.*

Proof. Let $x_0 \in U$ and $\overline{B}_{r_0}(x_0) \subset U$. We show that there is a $\delta > 0$ with $\overline{B}_\delta(F(x_0)) \subset F(U)$. Without loss of generality, we may assume $x_0 = 0$. Let $\delta > 0$ be such that

$$R_\delta = \inf\{r;\ \psi(\rho) > \delta \text{ for all } \rho > r\} < r_0.$$

Let $y \in \overline{B}_\delta(F(0))$ and let $F_n = F + \frac{1}{n}J$. It is clear that F_n is strongly accretive. Lemma 4.2 guarantees that the initial value problem

$$u' = -F_n(u) + y, \quad u(0) = x \in \overline{B}_{R_\delta}(0) \tag{4.1}$$

has a unique local solution $u(t; x)$. Let $\phi(t) = |u(t, x)|$. From 1), 2) and 4), we have

$$\phi(t)\, D^-\phi(t) \le \langle u', u\rangle_- \le \langle u', u\rangle_+$$

$$= \langle -F_n(u) + F_n(0) + y - F_n(0), u\rangle_+$$

$$\le -\langle F_n(u) - F_n(0), u\rangle_+ + |y - F_n(0)|\,\phi(t)$$

$$\le [|y - F(0)| - \psi(\phi(t))]\,\phi(t).$$

Hence
$$\phi(t)\, D^-\phi(t) \le [\delta - \psi(\phi(t))]\,\phi(t), \quad \phi(0) = |x|.$$

This implies that $\phi(t) \le R_\delta$. Consequently, $u(t, x)$ can be extended to a unique solution on $\mathbf{R}_+$ with $|u(t, x)| \le R_\delta$ for $t \ge 0$.

For each $t > 0$, consider the map

$$U(t) : \overline{B}_{R_\delta}(0) \to \overline{B}_{R_\delta}(0), \quad U(t)(x) = u(t; x).$$

Let us show that $U(t)$ is a contraction. Indeed, for $x_1, x_2 \in \overline{B}_{R_\delta}(0)$, let $v(t) = u(t; x_1) - u(t; x_2)$ and $\chi(t) = |v(t)|$. Then (4.1) implies

$$\chi(t)\, D^-\chi(t) \le \langle v'(t), v(t)\rangle_- \le \langle v'(t), v(t)\rangle_+$$

$$= -\langle F_n(u(t; x_1)) - F_n(u(t; x_2)), v(t)\rangle_+ \le -\frac{1}{n}\chi^2(t).$$

Integration yields $\ln \chi(t) - \ln \chi(0) \le -t/n$. Hence

$$\chi(t) \le \chi(0) \exp(-t/n).$$

Therefore,

$$|U(t)(x_1) - U(t)(x_2)| \le \rho_t |x_1 - x_2|,$$

where $\rho_t = \exp(-t/n) < 1$ for $t > 0$.

Also,

$$U(t)U(s) = U(s)U(t) = U(t+s).$$

Thus, $\{U(t); t > 0\}$ is a commuting family of contractions acting on $\overline{B}_{R_\delta}(0)$. It follows that there is a common fixed point $x_n \in \overline{B}_{R_\delta}(0)$, that is $u(t; x_n) = x_n$ for all $t > 0$. Consequently, $F_n(x_n) = y$. Hence,

$$F(x_n) = -\frac{1}{n}x_n + y \to y \quad \text{as} \quad n \to \infty.$$

Since

$$\psi(|x_n - x_m|) \le |F(x_n) - F(x_m)| \to 0 \quad \text{as } n, m \to \infty,$$

we have $x_n \to x$ for some $x \in \overline{B}_{R_\delta}(0)$ and $F(x) = y$. $\square$

Proposition 4.5 *Let X be a Banach space and $F : X \to X$ be ψ-accretive. Suppose that (a) or (b) holds. Then $F(X) = X$.*

Proof. By Lemma 4.3, $F(X)$ is closed, while by Lemma 4.4, $F(X)$ is open. Hence $F(X) = X$. $\square$

Proof of Theorem 4.1. As above, by replacing U with $U - x_0$ and $T(x)$ by $T(x + x_0)$, we may suppose $x_0 = 0$. Let

$$H : \overline{U} \times [0,1] \to X, \quad H(x, \lambda) = \lambda T(x).$$

It is clear that for each λ, there is a ψ_λ such that the map $J - H_\lambda = (1 - \lambda)J + \lambda(J - T)$ is ψ_λ-accretive. Let

$$\Lambda_H = \{\lambda \in [0,1]; H(x, \lambda) = x \text{ for some } x \in U\}.$$

Notice Λ_H is nonempty since $0 \in \Lambda_H$. The result will be established if we show that Λ_H is both closed and open in $[0,1]$.

We first show that Λ_H is closed. For this, let $(\lambda_k) \subset \Lambda_H$ be any sequence with $\lambda_k \to \lambda_0$ as $k \to \infty$. Since $\lambda_k \in \Lambda_H$, there is an $x_k \in U$

with $\lambda_k T(x_k) = x_k$. Notice that $(T(x_k))$ is bounded. Indeed, this is trivial if $T(\overline{U})$ is bounded. Otherwise, since

$$\psi(|x_k|)|x_k| \leq \langle x_k - T(x_k) + T(0), x_k \rangle_+$$

$$= -(1-\lambda_k)/\lambda_k |x_k|^2 + \langle T(0), x_k \rangle_+ \leq |T(0)||x_k|,$$

we have $\psi(|x_k|) \leq |T(0)|$. Now, since $\psi(r) \to \infty$ as $r \to \infty$, we see that (x_k) is bounded. Hence $(T(x_k))$ is bounded too. Thus $x_k - \lambda_0 T(x_k) \to 0$ as $k \to \infty$. As in the proof of Lemma 4.3, we find an $x_0 \in \overline{U}$ with $x_k \to x_0$ as $k \to \infty$ and $x_0 - \lambda_0 T(x_0) = 0$, which shows that $\lambda_0 \in \Lambda_H$.

Suppose now that Λ_H is not open in $[0,1]$. Then there exists $\lambda_0 \in \Lambda_H$ and a sequence $(\lambda_k) \subset [0,1]$ with $\lambda_k \notin \Lambda_H$ and $\lambda_k \to \lambda_0$ as $k \to \infty$. Let $x_0 \in U$ be such that $\lambda_0 T(x_0) = x_0$ and let $r > 0$ be such that $B = B_r(x_0) \subset U$. Then for each k,

$$y_k = (J - \lambda_k T)(x_0) \in (J - \lambda_k T)(B)$$

and $0 \notin (J - \lambda_k T)(B)$. From Lemma 4.4, $(J - \lambda_k T)(B)$ is open. So we may find

$$z_k \in \mathrm{seg}\,[0, y_k] \cap \partial(J - \lambda_k T)(B).$$

From Lemma 4.3, $(J - \lambda_k T)(\overline{B})$ is closed, so we have $\partial(J - \lambda_k T)(B) \subset (J - \lambda_k T)(\partial B)$. Thus, there exists $x_k \in \partial B$ with $(J - \lambda_k T)(x_k) = z_k$. Since $y_k \to 0$ as $k \to \infty$, we also have that $z_k \to 0$ as $k \to \infty$. Hence $(J - \lambda_0 T)(x_k) \to 0$ as $k \to \infty$, and again as in the proof of Lemma 4.3, we obtain an $\overline{x} \in \partial B$ with $x_k \to \overline{x}$ as $k \to \infty$ and $(J - \lambda_0 T)(\overline{x}) = 0$. Since $\overline{x} \neq x_0$, this contradicts the one-to-oneness of $J - \lambda_0 T$. $\square$

We now present a continuation theorem for accretive maps on Hilbert spaces which generalizes Theorem 3.5. The result is due to O'Regan [104].

Theorem 4.6 *Let X be a Hilbert and U an open bounded subset of X. Suppose $T : \overline{U} \to X$ is demicontinuous with $J - T$ accretive and $T(\overline{U})$ bounded. Also assume that*

$$(1-\lambda)x_0 + \lambda T(x) \neq x \quad \text{for all } x \in \partial U, \lambda \in [0,1]. \qquad (4.2)$$

for some $x_0 \in U$. Then T has a fixed point in U.

Proof. We may suppose $x_0 = 0$. For each $\mu \in (0,1)$, the map $T_\mu = -(1-\mu)J + T$ is demicontinuous, with $J - T_\mu$ strongly accretive. In addition, it is easy to check that T_μ also satisfies (4.2). Now Theorem 4.1 guarantees that there exists a unique $x_\mu \in U$ with $x_\mu - T_\mu(x_\mu) = 0$. Denote by x_k the element x_μ for $\mu = 1 - 1/(k-1)$. Then

$$x_k - (1 - 1/k)T(x_k) = 0.$$

As in the proof of Theorem 3.5, we have that (x_k) is convergent. It is clear that its limit is a fixed point of T. $\square$

4.3 Applications to Boundary Value Problems in Hilbert Spaces

We describe a typical example involving accretive maps.

Let us consider the two point boundary value problem

$$\begin{cases} u'' = f(t, u, u'), & t \in I = [0,1] \\ u(0) = u(1) = 0 \end{cases} \tag{4.3}$$

in a Hilbert space E. We look for classical solutions $u \in C^2(I; E)$ if f is continuous and Carathéodory solutions $u \in W^{2,2}(I; E)$ when f is L^2-Carathéodory.

We have the following existence results:

Theorem 4.7 *Let $(E, \langle ., . \rangle)$ be a Hilbert space and $f : I \times E \times E \to E$. Assume that f is continuous (resp., $f(., u, v)$ measurable for each $(u, v) \in E^2$ and $f(t, ., .)$ continuous for a.e. $t \in I$). In addition, suppose that the following conditions are satisfied:*

(i) *for each $r > 0$ there exists $h_r \in L^2(I)$ and $c_r \geq 0$ with*

$$|f(t, u, v)| \leq h_r(t) + c_r|v|, \quad a.e.\ t \in I, |u| \leq r, v \in E;$$

(ii) *there exists $a \geq 0$ and $b \geq 0$ with*

$$-\langle f(t, u, v) - f(t, \overline{u}, \overline{v}), u - \overline{u} \rangle \leq a|u - \overline{u}|^2 + b|v - \overline{v}||u - \overline{u}| \tag{4.4}$$

for a.e $t \in I$ and all $u, \overline{u}, v, \overline{v} \in E$;

(iii) $a/\pi^2 + b/4 < 1$.

Then (4.3) has a unique classical (resp., Carathéodory) solution.

Proof. We shall apply Proposition 4.5. Using the notation in Chapter 3, we note that (4.3) is equivalent to the fixed point problem

$$u = T(u), \quad u \in W_0^{1,2}(I; E)$$

for $T = L^{-1}Fj$, on the Hilbert space $W_0^{1,2}(I; E)$ endowed with the inner product

$$\langle u, v \rangle_{0,2} = \int_0^1 \langle u'(t), v'(t) \rangle \, dt$$

and the corresponding norm $\|u\|_{0,2} = \|u'\|_2$, where

$$j : W_0^{1,2}(I; E) \rightarrow C(I; E) \times L^2(I; E), \quad j(u) = (u, u'),$$

$$F : C(I; E) \times L^2(I; E) \rightarrow L^2(I; E), \quad F(u, v) = f(., u, v)$$

and

$$L^{-1} : L^2(I; E) \rightarrow W_0^{1,2}(I; E), \quad L^{-1}(v) = u, \quad u'' = v.$$

From (i), the map F is well defined, continuous and sends bounded sets into bounded sets. Also, j and L^{-1} are linear and bounded. Thus, T is continuous and sends bounded sets into bounded sets.

Now we show that $J-T$ is strongly accretive on $\left(W_0^{1,2}(I; E), \|\cdot\|_{0,2}\right)$. To do this we use the Wirtinger and Opial inequalities (Proposition 2.11). Note first, if $v \in L^2(I; E)$, then $u = L^{-1}(v)$ solves the problem $u''(t) = v(t)$ on I, $u(0) = u(1) = 0$. If we multiply the equation by $w(t)$, where $w \in W_0^{1,2}(I; E)$, and we integrate, we obtain

$$\left\langle L^{-1}v, w \right\rangle_{0,2} = -\langle v, w \rangle_2 \quad \text{for all } w \in W_0^{1,2}(I; E).$$

Now, for any $u_1, u_2 \in W_0^{1,2}(I; E)$, we have

$$\langle (J - T)(u_1) - (J - T)(u_2), u_1 - u_2 \rangle_{0,2}$$

$$= \|u_1 - u_2\|_{0,2}^2 - \langle T(u_1) - T(u_2), u_1 - u_2 \rangle_{0,2}$$

$$= \|u_1 - u_2\|_{0,2}^2 - \left\langle L^{-1}(F(u_1) - F(u_2)), u_1 - u_2 \right\rangle_{0,2}$$

$$= \|u_1 - u_2\|_{0,2}^2 + \langle F(u_1) - F(u_2), u_1 - u_2 \rangle_2.$$

Using (4.4), Wirtinger and Opial inequalities, we obtain

$$- \langle F(u_1) - F(u_2), u_1 - u_2 \rangle_2 \leq a \|u_1 - u_2\|_2^2 + b \|u_1 - u_2\|_{0,2}^2 / 4$$

$$\leq \left(\frac{a}{\pi^2} + \frac{b}{4} \right) \|u_1 - u_2\|_{0,2}^2.$$

Therefore

$$\langle (J - T)(u_1) - (J - T)(u_2), u_1 - u_2 \rangle_{0,2}$$

$$\geq \left(1 - \frac{a}{\pi^2} - \frac{b}{4} \right) \|u_1 - u_2\|_{0,2}^2.$$

Hence $J - T$ is strongly accretive and so all the assumptions of Proposition 4.5 are satisfied. $\square$

Theorem 4.8 *Suppose that all the assumptions of Theorem 4.7 are satisfied except (iii) which is relaxed to*

$$\frac{a}{\pi^2} + \frac{b}{4} \leq 1. \qquad\qquad (4.5)$$

In addition suppose that there exists a closed linear subspace X of $W_0^{1,2}(I; E)$ and a number $R > 0$ such that $T(X) \subset X$ and

$$\|u\|_{0,2} < R \qquad\qquad (4.6)$$

whenever $u \in X$ and $u = \lambda T(u)$ for some $\lambda \in [0,1]$. Then (4.3) has at least one classical (resp., Carathéodory) solution in X.

Proof. Using the argument in the proof of the above theorem, we see that T is continuous, sends bounded sets into bounded sets and $J - T$ is accretive. Thus we may apply Theorem 4.6, where U is the open ball $B_R(0)$ of $\left(X, \|\cdot\|_{0,2} \right)$ and $x_0 = 0$. $\square$

An a priori bound R like that in condition (4.6) can be obtained for nonlinearities f with symmetry:

Theorem 4.9 *Let $(E, \langle \cdot, \cdot \rangle)$ be a Hilbert space and $f : I \times E \times E \to E$ continuous. Assume that (i), (ii) and (4.5) hold. In addition suppose that f is symmetric, i.e.*

$$f(1 - t, u, -v) = f(t, u, v), \qquad\qquad (4.7)$$

and that there exists $r > 0$ *such that*

$$\langle u, f(t, u, v) \rangle + |v|^2 > 0$$

for all $t \in I$ *and all* $u, v \in E$ *satisfying* $|u| \geq r$ *and* $\langle u, v \rangle = 0$. *Then* (4.3) *has at least one classical solution* u *with* $u(t) = u(1 - t)$ *for all* $t \in I$.

Proof. We shall apply Theorem 4.8. Let

$$X = \left\{ u \in W_0^{1,2}(I; E); \ u(1 - t) = u(t) \text{ for all } t \in I \right\}.$$

For any $u \in X$, we have $u'(1 - t) = -u'(t)$ and, from (4.7), we obtain

$$(Fj)(u)(1 - t) = F(u, u')(1 - t)$$

$$= f(1 - t, u(1 - t), u'(1 - t)) = f(1 - t, u(t), -u'(t))$$

$$= f(t, u(t), u'(t)) = F(u, u')(t) = (Fj)(u)(t).$$

Also, if $v \in L^2(I; E)$ and $v(1 - t) = v(t)$, then direct computation using Green's function (2.32) shows that

$$\left(L^{-1}v\right)(1 - t) = \left(L^{-1}v\right)(t).$$

Consequently, $T(X) \subset X$. Now, let $u \in X$ with $u = \lambda T(u)$. Since f is continuous, $u \in C^2(I; E)$ and $u'(1/2) = 0$. As in the proof of Theorem 2.13, we have $\|u\|_\infty < r$. Then, from (i),

$$|u'(t)| = \left| \int_{1/2}^t u''(s)\, ds \right| \leq \int_{1/2}^t |f(s, u(s), u'(s))|\, ds$$

$$\leq \int_{1/2}^t (h_r(s) + c_r |u'(s)|)\, ds \leq A + c_r \int_{1/2}^t |u'(s)|\, ds$$

for all $t \in [1/2, 1]$ and some constant $A \geq 0$. Now Gronwall's inequality guarantees that $\|u'\|_\infty < R$ for some constant $R > 0$ independent of u. Clearly, $\|u\|_{0,2} = \|u'\|_2 \leq \|u'\|_\infty < R$. $\square$

5. Continuation Theorems Involving Compactness

This chapter presents Leray-Schauder type theorems for completely continuous maps on Banach spaces and their generalizations, maps of Krasnoselskii, Darbo, Sadovskii and Mönch type.

5.1 Mönch Continuation Principle

Let us start with the definitions.

Let X be a Banach space and $\mathcal{P}$ the family of all bounded sets of X. Denote by α the *Kuratowski measure of noncompactness* on X, that is $\alpha : \mathcal{P} \to \mathbf{R}_+$,

$$\alpha\left(S\right) = \inf \left\{d > 0; S \text{ admits a finite cover by sets of diameter } \leq d\right\}.$$

The following properties hold (for proof see [[32], Proposition 2.7.2]):
(a) $\alpha\left(S\right) = 0$ if and only if $\overline{S}$ is compact.
(b) α is a seminorm, i.e. $\alpha\left(\lambda S\right) = |\lambda|\,\alpha\left(S\right)$ and

$$\alpha\left(S_1 + S_2\right) \leq \alpha\left(S_1\right) + \alpha\left(S_2\right).$$

(c) $S_1 \subset S_2$ implies $\alpha\left(S_1\right) \leq \alpha\left(S_2\right)$. Also

$$\alpha\left(S_1 \cup S_2\right) = \max\left\{\alpha\left(S_1\right), \alpha\left(S_2\right)\right\}.$$

(d) $\alpha\left(\text{conv } S\right) = \alpha\left(S\right) = \alpha(\overline{S})$.

A continuous map $T : D \subset X \to X$ is said to be

(a) *compact* if $T(D)$ is relatively compact;

(b) *completely continuous* if $T(S)$ is relatively compact (i.e. $\overline{T(S)}$ is compact) for every bounded set $S \subset D$;

(c) of *Krasnoselskii type* if $T = T_1 + T_2$, where T_1 is completely continuous and T_2 is a contraction;

(d) *set-contraction* (of *Darbo type*) if there is a $\rho \in [0, 1)$ with

$$\alpha(T(S)) \leq \rho \alpha(S) \quad \text{for every } S \subset D \text{ bounded};$$

(e) *condensing* (of *Sadovskii type*) if

$$\alpha(T(S)) < \alpha(S) \quad \text{for every bounded } S \subset D \text{ with } \alpha(S) > 0;$$

(f) of *Mönch type* if there exists an $x_0 \in D$ with

$$S \subset D \text{ countable}, S \subset \overline{\text{conv}} \, \{\{x_0\} \cup T(S)\} \implies \overline{S} \text{ compact}.$$

Obviously, each compact map is completely continuous, each completely continuous map is of Krasnoselskii type; each map of Krasnoselskii type is a set-contraction; each set-contraction is condensing and each condensing map on a bounded set is of Mönch type. Also note that if D is bounded, then the notions of compact map and completely continuous map coincide.

Let us first recall Schauder's fixed point theorem.

Proposition 5.1 *Let X be a Banach space, $D \subset X$ nonempty closed convex and let $T : D \to D$ be a compact map. Then T has a fixed point.*

The following generalization due to Mönch [93] contains, as particular cases, the fixed point theorems of Krasnoselskii, Darbo and Sadovskii for self-maps of a closed bounded and convex set.

Proposition 5.2 *Let X be a Banach space, $D \subset X$ nonempty closed convex and $T : D \to D$ a continuous map satisfying*

$$S \subset D \text{ countable}, \overline{S} = \overline{\text{conv}} \, \{\{x_0\} \cup T(S)\} \implies \overline{S} \text{ compact} \quad (5.1)$$

for some $x_0 \in D$. Then T has a fixed point.

Proof. We construct the sequence

$$D_{k+1} = \text{conv} \left\{ \{x_0\} \cup T(D_k) \right\}, \quad D_0 = \{x_0\}.$$

Let $D' = \bigcup_{k \geq 0} D_k$ and $D^* = \overline{D'}$. We have that D' is a convex subset of D since $D_k \subset D_{k+1}$. In addition $D' = \text{conv} \{\{x_0\} \cup T(D')\}$. Hence D^* is closed convex and $T(D^*) \subset D^*$. On the other hand, we inductively see that $\overline{D}_k$ is compact. So there exists for each k a countable set $C_k \subset D_k$ with $\overline{C}_k = \overline{D}_k$. Consider the countable set $C = \bigcup_{k \geq 0} C_k$. We have $\overline{D'} = \overline{C}$. Also

$$\overline{\text{conv}} \left\{ \{x_0\} \cup T(C) \right\} = \overline{\text{conv}} \left\{ \{x_0\} \cup T(D') \right\} = \overline{C}.$$

Thus $\overline{C} = D^*$ is compact. So T is a continuous self-map of the nonempty compact convex set D^* and we may apply Schauder's fixed point theorem to deduce the result. $\square$

The corresponding Leray-Schauder type theorem is the following result also due to Mönch [93].

Theorem 5.3 *Let X be a Banach space, $K \subset X$ closed convex and $U \subset K$ bounded open in K. Denote by ∂U the boundary of U with respect to K. Assume $T : \overline{U} \to K$ is continuous and that for some $x_0 \in U$, the following condition is satisfied*

$$S \subset \overline{U} \text{ countable}, \ S \subset \overline{\text{conv}} \left\{ \{x_0\} \cup T(S) \right\} \implies \overline{S} \text{ compact}. \quad (5.2)$$

In addition assume that

$$(1 - \lambda) x_0 + \lambda T(x) \neq x \quad \text{for all } x \in \partial U \text{ and } \lambda \in [0, 1]. \quad (5.3)$$

Then T has a fixed point in U.

Proof. If $U = K$, then the conclusion follows directly from Proposition 5.2. Assume $U \neq K$ and so that $\partial U \neq \emptyset$. Define the homotopy $H : \overline{U} \times [0,1] \to K$ by

$$H(x, \lambda) = (1 - \lambda) x_0 + \lambda T(x) \quad (5.4)$$

and let

$$\Sigma = \{ x \in \overline{U}; \ H(x, \lambda) = x \text{ for some } \lambda \in [0, 1] \}.$$

Since H is continuous, Σ is closed. Also (5.3) guarantees that Σ and ∂U are disjoint. So, from Urysohn's lemma, there exists $v \in C(\overline{U}; [0,1])$ with $v(x) = 0$ on ∂U and $v(x) = 1$ on Σ. Define

$$D = \overline{\text{conv}}\,\{\{x_0\} \cup T(\overline{U})\}$$

and $\widehat{T} : D \to D$ by

$$\widehat{T}(x) = \begin{cases} H(x, v(x)) & \text{for } x \in \overline{U} \\ x_0 & \text{for } x \notin \overline{U}. \end{cases}$$

It is easy to check that $\widehat{T}$ is a continuous self-map of D. Now we prove that $\widehat{T}$ satisfies (5.1). Let $S \subset D$ be a countable set with $\overline{S} = \overline{\text{conv}}\,\{\{x_0\} \cup \widehat{T}(S)\}$. Using (5.4), we see that

$$\overline{\text{conv}}\,\{\{x_0\} \cup \widehat{T}(S)\} = \overline{\text{conv}}\,\{\{x_0\} \cup T(S \cap \overline{U})\}.$$

So, by (5.2), $S \cap \overline{U}$ is relatively compact. Further the continuity of T implies that $T(S \cap \overline{U})$ is relatively compact. Thus, by Mazur's lemma, the entire set $\overline{S}$ is compact. Therefore, we may apply Proposition 5.2 to $\widehat{T}$ and we may deduce that there exists an $x \in D$ which is a fixed point of $\widehat{T}$. Since $x_0 \in U$, we have $x \in \overline{U}$ and so $H(x, v(x)) = x$. This shows that $x \in \Sigma$. Consequently, $v(x) = 1$ and so $T(x) = x$. $\square$

Remark 5.1 *In particular, if T is condensing, set-contraction, or compact, then so also is $\widehat{T}$ and so instead of Proposition 5.2 we may use the fixed point theorem of Sadovskii, Darbo, or Schauder, respectively. Thus, Theorem 5.3 becomes in particular, a theorem of Leray-Schauder type for condensing, set-contraction and compact maps.*

5.2   Granas' Topological Transversality Theorem

In Theorem 5.3, the map T is connected to the constant map x_0 by means of the simplest homotopy $(1 - \lambda) x_0 + \lambda T$. The question we ask is what kind of maps can replace the constant map x_0 so that the conclusion of Theorem 5.3 remain valid for other homotopies? For an answer, we need the notion of an essential map.

Let X be a Banach space, $K \subset X$ closed convex and $U \subset K$ nonempty bounded and open in K. Let $\mathbf{M}$ be any of the following classes $\mathbf{M}_S$, $\mathbf{M}_D$, $\mathbf{M}_C$ of all condensing, set-contractions, respectively compact maps from $\overline{U}$ into K, which are fixed point free on ∂U. In the next chapter we shall write $\mathbf{M}(\overline{U}; K)$ instead of $\mathbf{M}$, when this will be important. A map $T \in \mathbf{M}$ is said to be *essential in* $\mathbf{M}$ if every $T' \in \mathbf{M}$ with $T'(x) = T(x)$ on ∂U has a fixed point.

The following result is known as the *topological transversality theorem* and is essentially due to Granas [55] (see also [154], [51], [76], [127]).

Theorem 5.4 *Let* $H : \overline{U} \times [0,1] \to K$ *be condensing (set-contraction, compact, respectively). Assume*

(a) $H(x, \lambda) \neq x$ *for all* $x \in \partial U$ *and* $\lambda \in [0,1]$;

(b) H_0 *is essential in* $\mathbf{M}_S$ ($\mathbf{M}_D$, $\mathbf{M}_C$, *respectively).*

Then, for each $\lambda \in [0,1]$, *there exists a fixed point of* H_λ *in* U. *Moreover,* H_λ *is essential in* $\mathbf{M}_S$ ($\mathbf{M}_D$, $\mathbf{M}_C$, *respectively) for every* $\lambda \in [0,1]$.

Proof. It is sufficient to prove that the conclusion holds for $\lambda = 1$. Let $T_1 \in \mathbf{M}$ be any map with $T_1(x) = H_1(x)$ on ∂U. Define a new homotopy connecting H_0 to T_1 via H_1,

$$
\widehat{H}(x, \lambda) = \begin{cases} H(x, 2\lambda), & \lambda \in [0, 1/2] \\[2mm] 2(1-\lambda) H_1(x) + 2(\lambda - 1/2) T_1(x), & \lambda \in [1/2, 1] \end{cases}
$$

It is easy to see that $\widehat{H}$ is condensing (respectively, set-contraction or compact) like H. Let

$$
\Sigma = \{x \in \overline{U}; \quad \widehat{H}(x, \lambda) = x \text{ for some } \lambda \in [0,1]\}.
$$

Clearly Σ and ∂U are disjoint. Let $v \in C(\overline{U}; [0,1])$ with $v(x) = 0$ on ∂U and $v(x) = 1$ on Σ. The map $T_0(x) = \widehat{H}(x, v(x))$ belongs to $\mathbf{M}$ and $T_0(x) = H_0(x)$ on ∂U. Since H_0 is essential, T_0 has a fixed point, say y. Then $y \in \Sigma$ and so $v(y) = 1$. Consequently, $T_0(y) = T_1(y)$ which shows that y is a fixed point of T_1. $\square$

Proposition 5.5 *Any constant map* x_0, *where* $x_0 \in U$, *is essential in* $\mathbf{M}_S$ ($\mathbf{M}_D$, $\mathbf{M}_C$, *respectively).*

Proof. Let $T \in \mathbf{M}$ and $T(x) = x_0$ on ∂U. The set $D = \overline{\text{conv }} T(\overline{U})$ is nonempty $(x_0 \in D)$ closed bounded and convex, while the operator

$$\widehat{T}(x) = \begin{cases} T(x) \text{ for } x \in \overline{U} \\ x_0 \text{ for } x \in D \setminus \overline{U} \end{cases}$$

is condensing (respectively, set-contraction or compact) and maps D into itself. Consequently, by Sadovskii (respectively, Darbo, Schauder) fixed point theorem, there exists $y \in D$ with $\widehat{T}(y) = y$. We now observe that $y \notin D \setminus \overline{U}$ and hence $\widehat{T}(y) = T(y)$. Thus T has a fixed point. □

5.3   Measures of Noncompactness on C(I;E)

In this section we derive some useful results for equicontinuous subsets of $C(I; E)$, where E is a Banach space.

For a family M of functions $u : I \to E$, we define $M(t) = \{u(t)\,;\ u \in M\} \subset E$ for each $t \in I$.

Lemma 5.6 *Let* $(E, |\,.\,|)$ *be a Banach space and* $M \subset C(I; E)$ *equicontinuous with* $M(t)$ *bounded for each* $t \in I$. *Then*

1) *for all* $t, s \in I$, *one has*

$$|\alpha(M(t)) - \alpha(M(s))| \le 2\omega(M; |t - s|),$$

where $\omega(M; \delta)$ *is the modulus of continuity of* M, *namely*

$$\omega(M; \delta) = \sup\{|u(t) - u(s)|\,;\ t, s \in I, |t - s| \le \delta, u \in M\}.$$

2) *We have*

$$\alpha\left(\int_0^1 M(t)\,dt\right) \le \int_0^1 \alpha(M(t))\,dt, \tag{5.5}$$

where

$$\int_0^1 M(t)\,dt = \left\{\int_0^1 u(t)\,dt;\ u \in M\right\}.$$

.

Proof. 0) We first prove that for every bounded sets $S_1, S_2 \subset E$,

$$|\alpha(S_1) - \alpha(S_2)| \le 2\,d_{HP}(S_1, S_2), \tag{5.6}$$

where d_{HP} denotes the *Hausdorff-Pompeiu metric*, i.e.

$$d_{HP}(S_1, S_2) = \max\{\sup_{x \in S_1} d(x, S_2),\ \sup_{x \in S_2} d(x, S_1)\},$$

here d being the distance from an element of E to a set of E, i.e.
$d(x, S) = \inf_{y \in S} |x - y|$.

To prove (5.6), let $\varepsilon > 0$ and take a partition of S_1, $S_1 = \bigcup\limits_{k=1}^{m} S_{1k}$
with diam $S_{1k} < \alpha(S_1) + \varepsilon$ for all k. Let $\mu = d_{HP}(S_1, S_2) + \varepsilon$ and

$$S_{2k} = \{y \in S_2;\ \text{there is } x \in S_{1k} \text{ with } |x - y| < \mu\},$$

$k = 1, 2, ..., m$. It is easy to see that $S_2 = \bigcup\limits_{k=1}^{m} S_{2k}$ and

$$\text{diam} S_{2k} \leq \text{diam} S_{1k} + 2\mu < \alpha(S_1) + 2\,d_{HP}(S_1, S_2) + 3\varepsilon.$$

Hence

$$\alpha(S_2) < \alpha(S_1) + 2\,d_{HP}(S_1, S_2) + 3\varepsilon.$$

Similarly

$$\alpha(S_1) < \alpha(S_2) + 2\,d_{HP}(S_1, S_2) + 3\varepsilon.$$

Thus

$$|\alpha(S_1) - \alpha(S_2)| < 2\,d_{HP}(S_1, S_2) + 3\varepsilon.$$

Since ε is arbitrary, this implies (5.6).

1) Now the first part of Lemma 5.6 immediately follows using the obvious inequality

$$d_{HP}(M(t), M(s)) \leq \omega(M; |t - s|).$$

2) By the equicontinuity of M, for each $\varepsilon > 0$, there is $\delta > 0$ such that $|u(t) - u(t')| < \varepsilon$ for $|t - t'| < \delta$ and all $u \in M$. Thus, $d_{HP}(M(t), M(t')) \leq \varepsilon$ whenever $|t - t'| < \delta$. Consequently,

$$|\alpha(M(t)) - \alpha(M(t'))| \leq 2\varepsilon \quad \text{for } |t - t'| < \delta.$$

Thus $\alpha(M(t))$ is continuous on I. To prove (5.5), for any integer m, $m \geq 1$, consider the partition $(t_k)_{k=1}^{m}$ of I, where $t_0 = 0$, $t_m = 1$ and $t_k - t_{k-1} = 1/m$ for $1 \leq k \leq m$. For any given $\varepsilon > 0$, there is N such that $m > N$ implies

$$|u(t_k) - u(t)| < \varepsilon \quad \text{for all } u \in M,\ t \in [t_{k-1}, t_k], 1 \leq k \leq m.$$

Then

$$\left| \sum_{k=1}^{m} u\left(t_{k}\right)/m - \int_{0}^{1} u\left(t\right) dt \right| = \left| \sum_{k=1}^{m} \int_{t_{k-1}}^{t_{k}} \left[u\left(t_{k}\right) - u\left(t\right)\right] dt \right|$$

$$\leq \sum_{k=1}^{m} \int_{t_{k-1}}^{t_{k}} \left| u\left(t_{k}\right) - u\left(t\right) \right| dt < \varepsilon, \quad u \in M, \ m > N.$$

It follows that

$$d_{HP}\left(\left\{ \sum_{k=1}^{m} u\left(t_{k}\right)/m; \ u \in M \right\}, \int_{0}^{1} M\left(t\right) dt\right) \leq \varepsilon, \quad m > N$$

and by (5.6),

$$\left| \alpha\left(\left\{ \sum_{k=1}^{m} u\left(t_{k}\right)/m; \ u \in M \right\}\right) - \alpha\left(\int_{0}^{1} M\left(t\right) dt\right) \right| \leq 2\varepsilon.$$

Consequently

$$\alpha\left(\left\{ \sum_{k=1}^{m} u\left(t_{k}\right)/m; \ u \in M \right\}\right) \to \alpha\left(\int_{0}^{1} M\left(t\right) dt\right) \quad \text{as } m \to \infty.$$

On the other hand

$$\alpha\left(\left\{ \sum_{k=1}^{m} u\left(t_{k}\right)/m; u \in M \right\}\right) \leq \frac{1}{m} \sum_{k=1}^{m} \alpha\left(M\left(t_{k}\right)\right) \to \int_{0}^{1} \alpha\left(M\left(t\right)\right) dt.$$

□

The next result is a generalization due to Ambrosetti [8] of the classical Arzelà-Ascoli theorem.

Lemma 5.7 *Let E be a Banach space and $M \subset C\left(I; E\right)$ equicontinuous with $M\left(t\right)$ bounded for each $t \in I$. Then M is bounded in $C\left(I; E\right)$ and*

$$\alpha\left(M\right) = \max \left\{ \alpha\left(M\left(t\right)\right); t \in I \right\}.$$

Proof. We first observe that, by Lemma 5.6, $\alpha\left(M\left(t\right)\right)$ is a continuous function on I, so $\max\left\{\alpha\left(M\left(t\right)\right);\ t\in I\right\}$ exists.

For a given $\varepsilon>0$, by the equicontinuity of M, we can cover I by a finite number of subintervals $I\left(t_k\right)$, $k=1,\ 2,\ ...,\ m$, with

$$\left|u\left(t\right)-u\left(t_k\right)\right|<\varepsilon \quad \text{for all } u\in M,\ t\in I\left(t_k\right) \text{ and } k. \qquad (5.7)$$

For any $t\in I$, there is a $k\in\left\{1,\ 2,\ ...,\ m\right\}$ with $t\in I\left(t_k\right)$. Then

$$\left|u\left(t\right)\right|\leq\left|u\left(t\right)-u\left(t_k\right)\right|+\left|u\left(t_k\right)\right|\leq\varepsilon+C_k,$$

where C_k is a bound of $\left|M\left(t_k\right)\right|$. Consequently,

$$\left|u\left(t\right)\right|\leq\varepsilon+C \quad \text{for all } t\in I,$$

where $C=\max\left\{C_k;\ k=1,\ 2,\ ...,\ m\right\}$, which shows that M is bounded and so, we may speak about $\alpha\left(M\right)$.

Let $\mu>\max\left\{\alpha\left(M\left(t\right)\right);\ t\in I\right\}$. From (5.7) we have

$$M\left(t\right)\subset\bigcup_{k=1}^{m}\left(M\left(t_k\right)+B_\varepsilon\left(0\right)\right) \quad \text{for all } t\in I.$$

Then, there exist $E_1,\ E_2,\ ...,\ E_p\subset E$ with

$$\operatorname{diam}E_j\leq\mu \quad \text{and} \quad \bigcup_{k=1}^{m}M\left(t_k\right)\subset\bigcup_{j=1}^{p}E_j.$$

Thus, M is the union of the finitely many sets

$$\left\{u\in M;\ u\left(t_1\right)\in E_{j_1},u\left(t_2\right)\in E_{j_2},...,u\left(t_m\right)\in E_{j_m}\right\},$$

each of which has diameter $\leq\mu+2\varepsilon$. Hence $\alpha\left(M\right)\leq\mu+2\varepsilon$ and as a result,

$$\alpha\left(M\right)\leq\max\left\{\alpha\left(M\left(t\right)\right);\ t\in I\right\}.$$

To prove the opposite inequality, let $\mu>\alpha\left(M\right)$. Then there is a partition $M=\bigcup_{k=1}^{p}M_k$ with $\operatorname{diam}M_k\leq\mu$. Hence, for any $t\in I$,

$M\left(t\right)=\bigcup_{k=1}^{p}M_k\left(t\right)$ and $\operatorname{diam}M_k\left(t\right)\leq\operatorname{diam}M_k\leq\mu$. Consequently,

$$\max\left\{\alpha\left(M\left(t\right)\right);\ t\in I\right\}\leq\alpha\left(M\right).$$

$\square$

From Lemma 5.7, it follows that a set $M \subset C(I;E)$ is relatively compact if and only if it is equicontinuous and $M(t)$ is relatively compact for each $t \in I$.

We finish this section with some properties related to another natural measure of noncompactness on a Banach space X, namely the *ball measure of noncompactness*: $\alpha^\circ : \mathcal{P} \to \mathbf{R}_+$,

$$\alpha^\circ(S) = \inf\{r > 0; S \text{ admits a finite cover by balls of radius } r\}.$$

Notice that the properties (a)-(d) of α are also true for α°. In addition,

$$\alpha^\circ(S) \leq \alpha(S) \leq 2\alpha^\circ(S) \quad \text{for all } S \in \mathcal{P}.$$

Here are two interesting properties of α° (see [93]):

1) Let X be a separable Banach space, $\{x_k; k \geq 1\}$ dense in X and let X_n be the subspace generating by $\{x_k; 1 \leq k \leq n\}$. The ball measure of noncompacness of a countable subset $S = \{y_k; k \geq 1\} \subset X$ can be computed from the formula

$$\alpha^\circ(S) = \lim_{n \to \infty} \overline{\lim_{m \to \infty}} \, d(y_m, X_n).$$

2) If E is a separable Banach space, $M \subset C(I;E)$ is countable and bounded, then the function $t \longmapsto \alpha^\circ(M(t))$ is measurable and

$$\alpha^\circ\left(\int_0^1 M(t)\, dt\right) \leq \int_0^1 \alpha^\circ(M(t))\, dt.$$

5.4   The Cauchy Problem in Banach Spaces

In this section we shall generalize some of the existence results described in Chapter 2 for the Cauchy problem.

Consider the Cauchy problem

$$\begin{cases} u' = f(t,u), & t \in I \\ u(0) = u_0 \end{cases} \tag{5.8}$$

and the related parametrized family of problems

$$\begin{cases} u' = \lambda f(t,u), & t \in I \\ u(0) = u_0, \end{cases} \tag{5.9}$$

where $\lambda \in [0,1]$.

We seek a *classical solution*, that is $u \in C^1(I;E)$ if f is continuous, and a *weak solution* $u \in W^{1,1}(I;E)$ when f is only Carathéodory.

There are many papers concerning the Cauchy problem in Banach spaces. We refer the reader to [6], [11], [31], [87] and the references therein for a history and additional aspects of this problem.

Suppose $w : I \times \mathbf{R}_+ \to \mathbf{R}_+$ is continuous. We say that w is a *Kamke function* of the Cauchy problem if the unique solution $u \in C(I;\mathbf{R}_+)$ to the integral inequality

$$u(t) \leq \int_0^t w(\tau, u(\tau)) \, d\tau, \quad t \in I,$$

which satisfies $u(0) = 0$, is $u(t) \equiv 0$.

We have the following existence principle which combines well known earlier results (see [11], [46] and [94]).

Theorem 5.8 *Let E be a Banach space, $u_0 \in E$, $R > 0$ and let $f : I \times \overline{B}_R(u_0; E) \to E$ be a Carathéodory function. Assume $f = g + h$ where g is uniformly continuous and h is a Carathéodory function. Also assume that there exists a Kamke function w, $\eta \in L^1(I;\mathbf{R}_+)$ and a compact set $\Psi \subset E$ such that for each $S \subset \overline{B}_R(u_0; E)$,*

$$\alpha(g(t,S)) \leq w(t, \alpha(S)) \quad \text{for all } t \in I \tag{5.10}$$

and

$$h(t,u) \in \eta(t)\Psi \quad \text{for } |u - u_0| \leq R \text{ and a.e. } t \in I. \tag{5.11}$$

In addition assume that

(B) *for every solution $u \in W^{1,1}(I;E)$ to (5.9), where $\lambda \in (0,1]$, one has $|u(t) - u_0| < R$ for all $t \in I$.*

Then there exists $u \in W^{1,1}(I;E)$ that solves (5.8).

Proof.   Clearly, $u \in W^{1,1}(I;E)$ solves (5.8) if and only if $u \in C(I;E)$ solves the integral equation

$$u(t) = u_0 + \int_0^t f(\tau, u(\tau)) \, d\tau, \quad t \in I,$$

equivalently, if u is a fixed point of the continuous map $T : D \to C$, where $C = C(I; E)$, $D = \{u \in C; \ |u(t) - u_0| \leq R \text{ on } I\}$ and

$$T(u)(t) = u_0 + \int_0^t f(\tau, u(\tau)) \, d\tau. \tag{5.12}$$

We apply Theorem 5.3 to $X = C$, $x_0 = u_0$,

$$K = \overline{\text{conv}} \ \{\{u_0\} \cup T(D)\}$$

and

$$U = \{u \in K; \ |u(t) - u_0| < R \text{ on } I\}.$$

Obviously, condition (B) implies (5.3). Now we check (5.2). First we note that, since D is bounded and f is a Carathéodory function, $T(D)$ is bounded and equicontinuous. Consequently, K is bounded and equicontinuous too. Hence each part of K is bounded and equicontinuous. This is the reason for this choice of K. Now let $S \subset \overline{U}$ be such that $S \subset \overline{\text{conv}} \ \{\{u_0\} \cup T(S)\}$. For each $t \in I$, $S(t) \subset \overline{\text{conv}} \ \{\{u_0\} \cup T(S)(t)\}$. It follows that $\alpha(S(t)) \leq \alpha(T(S)(t))$. Since g is uniformly continuous and S is equicontinuous, we have that $g(.,S(.))$ is equicontinuous too. In addition, from (5.10), it follows that the set $g(t, S(t))$ is bounded for each t. Using Lemma 5.6 we obtain

$$\alpha \left(u_0 + \int_0^t g(\tau, S(\tau)) \, d\tau \right) = \alpha \left(\int_0^t g(\tau, S(\tau)) \, d\tau \right) \tag{5.13}$$

$$\leq \int_0^t \alpha(g(\tau, S(\tau))) \, d\tau \leq \int_0^t w(\tau, \alpha(S(\tau))) \, d\tau.$$

Unfortunately, we can not prove that

$$\alpha \left(\int_0^t h(\tau, S(\tau)) \, d\tau \right) \leq \int_0^t \alpha(h(\tau, S(\tau))) \, d\tau.$$

We can avoid this step by using a trick from [46] as follows: Let $b^* \in E^*$ and suppose Ψ is contained in the half-space where $b^* \leq c$, that is $b^*(x) \leq c$ for all $x \in \Psi$. From (5.11), for each $u \in S$ and almost all $\tau \in I$,

$$h(\tau, u(\tau)) = \eta(\tau) k_{u,\tau} \quad \text{for some } k_{u,\tau} \in \Psi.$$

Then

$$b^*(h(\tau, u(\tau))) = \eta(\tau) b^*(k_{u,\tau}) \leq c\eta(\tau).$$

Integration gives

$$b^* \left(\int_0^t h\left(\tau, u\left(\tau\right)\right) d\tau \right) \leq c \int_0^t \eta\left(\tau\right) d\tau.$$

Since the intersection of all half-spaces that contain Ψ is its closed convex hull, we find that

$$\int_0^t h\left(\tau, u\left(\tau\right)\right) d\tau \in \left(\int_0^t \eta\left(\tau\right) d\tau \right) \overline{\mathrm{conv}}\, \Psi.$$

Now Ψ is compact and so we have

$$\alpha \left(\int_0^t h\left(\tau, S\left(\tau\right)\right) d\tau \right) = 0.$$

Thus

$$\alpha\left(S\left(t\right)\right) \leq \alpha\left(T\left(S\right)\left(t\right)\right) = \alpha \left(\int_0^t (g+h)\left(\tau, S\left(\tau\right)\right) d\tau \right)$$

$$\leq \alpha \left(\int_0^t g\left(\tau, S\left(\tau\right)\right) d\tau \right) \leq \int_0^t w\left(\tau, \alpha\left(S\left(\tau\right)\right)\right) d\tau.$$

Hence

$$\alpha\left(S\left(t\right)\right) \leq \int_0^t w\left(\tau, \alpha\left(S\left(\tau\right)\right)\right) d\tau, \quad \alpha\left(S\left(0\right)\right) = 0.$$

Thus $\alpha\left(S\left(t\right)\right) = 0$ for all $t \in I$ and from Lemma 5.7, $\alpha\left(S\right) = 0$. Therefore $\overline{S}$ is compact. $\square$

Remark 5.2 *In the special case when* $w\left(t, u\right) = l\left(t\right)u$, *where* $l \in C\left(I; \mathbf{R}_+\right)$, *and*

$$\alpha\left(g\left(t, S\right)\right) \leq l\left(t\right)\alpha\left(S\right) \quad \text{for any } S \subset \overline{B}_R\left(u_0; E\right), \ t \in I, \qquad (5.14)$$

the map $T : \overline{U} \to X$ *given by (5.12) is an* α_θ-*set-contraction. Here* α_θ *is the Kuratowski measure of noncompactness with respect to the Bielecki norm on* C,

$$\|u\|_\theta = \max_{t \in I} \left[\exp\left(-\theta \int_0^t l\left(s\right) ds\right) |u\left(t\right)| \right],$$

with $\theta > 1$. *Indeed, for each* $M \subset \overline{U}$ *(consequently, bounded and equicontinuous), we have*

$$\alpha\left(T\left(M\right)\left(t\right)\right) = \alpha \left(\int_0^t f\left(\tau, M\left(\tau\right)\right) d\tau \right) \leq \int_0^t \alpha\left(g\left(\tau, M\left(\tau\right)\right)\right) d\tau$$

$$\leq \int_0^t l\left(\tau\right) \alpha\left(M\left(\tau\right)\right) d\tau \; \leq \; \alpha_\theta\left(M\right) \int_0^t l\left(\tau\right) \exp(\theta \int_0^\tau l\left(s\right) ds) d\tau$$

$$\leq \frac{1}{\theta} \alpha_\theta\left(M\right) \exp(\theta \int_0^t l\left(s\right) ds).$$

Dividing by $\exp(\theta \int_0^t l\left(s\right) ds)$ *and taking maximum when* $t \in I$ *we obtain*

$$\alpha_\theta\left(T\left(M\right)\right) \leq \frac{1}{\theta} \alpha_\theta\left(M\right).$$

Thus, under assumption (5.14), the result in Theorem 5.8 follows from the Leray-Schauder principle for set-contractions.

Remark 5.3 *If* $g\left(t, u\right)$ *is Lipschitz with respect to* u *with Lipschitz constant* l, *for all* $t \in I$, *then (5.14) holds and the map* T *is the sum of a contraction with respect to the norm* $\|.\|_\theta$ *and a completely continuous map, that is a Krasnoselskii map.*

Remark 5.4 *If* $g \equiv 0$, *then* T *is compact and the result follows from the Leray-Schauder principle for compact maps. This case was considered in [46].*

Remark 5.5 *If* $h \equiv 0$, *then the solutions of (5.8) are classical.*

Theorem 5.8 implies the following existence result.

Theorem 5.9 *Let* E *be a Banach space and* $f : I \times E \to E$ *a Carathéodory function. Assume* $f = g + h$, *where* g *is uniformly continuous on bounded sets and* h *is a Carathéodory function. In addition suppose that the following conditions are satisfied:*

(i) *for each* $R > 0$, *there exists a Kamke function* w_R, $\eta_R \in L^1\left(I; \mathbf{R}_+\right)$ *and* $\Psi_R \subset E$ *compact such that*

$$\alpha\left(g\left(t, S\right)\right) \leq w_R\left(t, \alpha\left(S\right)\right)$$

for all $t \in I$ *and every* $S \subset \overline{B}_R\left(u_0; E\right)$, *and*

$$h\left(t, u\right) \in \eta_R\left(t\right) \Psi_R$$

for $|u - u_0| \leq R$ *and a.e.* $t \in I$;

(ii) *there exists* $\beta \in L^1 (I; \mathbf{R}_+)$ *and* $\psi : \mathbf{R}_+ \to (0, \infty)$ *nondecreasing with* $1/\psi \in L^1_{\text{loc}} (\mathbf{R}_+; \mathbf{R})$ *such that*

$$|f(t, u)| \leq \beta(t) \psi(|u|) \quad \text{for all } u \in E \text{ and a.e. } t \in I$$

and

$$\int_{|u_0|}^{\infty} \frac{d\tau}{\psi(\tau)} > \int_0^1 \beta(\tau) \, d\tau.$$

Then there exists $u \in W^{1,1} (I; E)$ *that solves (5.8).*

Proof. As in the proof of Theorem 2.6 we can show that there exists $R > 0$ such that $|u(t) - u_0| < R$ for all $t \in I$ and any solution $u \in W^{1,1} (I; E)$ to (5.9). Then we apply Theorem 5.8. $\square$

Remark 5.6 *Remark 2.3 gives us an additional result when* E *is a Hilbert space.*

Remark 5.7 *A variant of Theorem 5.8 can be stated assuming that* g *is only continuous and bounded, and that condition (5.10) holds for countable sets* S, *with* α° *instead of* α. *The idea is to consider a separable closed subspace* E_S *of* E *such that* $g(t, S(t)) \subset E_S$ *for all* $t \in I$, *and to use property 2) of* α° *from the end of the previous section (see [94]).*

5.5 Sturm-Liouville Problems in Banach Spaces

We now return to the boundary value problem

$$\begin{cases} u'' = f(t, u, u'), & t \in I \\ u \in \mathcal{B} \end{cases} \tag{5.15}$$

in a Banach space E, where $\mathcal{B}$ represents the homogenous Sturm-Liouville boundary conditions

$$\begin{cases} u(0) - au'(0) = 0 \\ u(1) + bu'(1) = 0 \end{cases}$$

with $a, b \geq 0$. Then the linear operator $L : C^1_\mathcal{B} \to C$, $Lu = u''$ is invertible and we denote by $G(t, s)$ the corresponding Green function. The inverse of L can be extended to $L^1 (I; E)$ as follows

$$L^{-1} : L^1 (I; E) \to W^{2,1} (I; E) \subset C^1,$$

$$(L^{-1}v)(t) = \int_0^1 G(t,\tau)\, v(\tau)\, d\tau.$$

We seek weak solutions $u \in W^{2,1}(I;E)$ if $f(t,z)$, $z = (u,v)$, is a Carathéodory function, and classical solutions $u \in C^2(I;E)$ if f is continuous. Let us also consider the related parametrized family of problems

$$\begin{cases} u'' = \lambda f(t,u,u'), & t \in I \\ u \in \mathcal{B} \end{cases} \qquad (5.16)$$

where $\lambda \in [0,1]$.

We say that a continuous function $w : I \times \mathbf{R}_+^2 \to \mathbf{R}_+$ is a *Kamke function of the Sturm-Liouville problem* if the unique solution (φ, ψ), $\varphi, \psi \in C(I;\mathbf{R}_+)$, of the following system of inequations

$$\varphi(t) \leq \int_0^1 |G(t,\tau)|\, w(\tau, \varphi(\tau), \psi(\tau))\, d\tau \qquad (5.17)$$

$$\psi(t) \leq \int_0^1 |G_t(t,\tau)|\, w(\tau, \varphi(\tau), \psi(\tau))\, d\tau$$

satisfying $\varphi(0) - a\psi(0) = 0$ and $\varphi(1) - b\psi(1) = 0$, is $\varphi \equiv 0$, $\psi \equiv 0$.

Theorem 5.3 yields the following existence principle.

Theorem 5.10 *Let E be a Banach space, $R > 0$ and $f : I \times \overline{B}_R(0;E)^2 \to E$ a Carathéodory function. Assume $f = g + h$ where g is uniformly continuous and h is a Carathéodory function. Also suppose that the following conditions are satisfied:*

(a) *there exists a Kamke function w such that for each $S = (S_1, S_2) \subset \overline{B}_R(0;E)^2$,*

$$\alpha(g(t,S)) \leq w(t, \alpha(S_1), \alpha(S_2)) \quad \text{for all } t \in I;$$

(b) *there exists $\eta \in L^1(I;\mathbf{R}_+)$ and $\Psi \subset E$ compact such that*

$$h(t,u,v) \in \eta(t)\,\Psi \quad \text{for } |u| \leq R,\ |v| \leq R \text{ and a.e. } t \in I;$$

(c) *for every solution $u \in W^{2,1}(I;E)$ to (5.16), where $\lambda \in [0,1]$, one has*

$$|u(t)| < R,\ |u'(t)| < R \quad \text{for all } t \in I.$$

Then there exists $u \in W^{2,1}(I;E)$ that solves (5.15).

Proof. A function $u \in W^{2,1}(I; E)$ solves (5.15) if $u \in C_{\mathcal{B}}^1$ solves the integral equation

$$u(t) = \int_0^1 G(t, \tau) f(\tau, u(\tau), u'(\tau)) \, d\tau, \quad t \in I$$

equivalently, $u = L^{-1} N_f(u)$, where

$$N_f : D_0 \to L^1(I; E), \quad N_f(u)(t) = f(t, u(t), u'(t)),$$

$$D_0 = \left\{ u \in C_{\mathcal{B}}^1; \, |u(t)| \le R, \, |u'(t)| \le R \text{ for all } t \in I \right\}.$$

Hence u is a fixed point of the continuous map $T : D_0 \to C_{\mathcal{B}}^1$, $T = L^{-1} N_f$.

We apply Theorem 5.3 with $X = C_{\mathcal{B}}^1$ and norm $\|u\| = \max \{\|u\|_\infty, \|u'\|_\infty\}$ on C^1, $K = \overline{\text{conv}} \{\{0\} \cup T(D_0)\}$, x_0 the null function and $U = \{u \in K; \|u\| < R\}$. Now (c) guarantees that (5.3) holds. We now check (5.2). Notice that both K and K' are equicontinuous. Let $S \subset \overline{U}$ be such that $S \subset \overline{\text{conv}} \{\{0\} \cup T(S)\}$. Let

$$\varphi(t) = \alpha(S(t)) \quad \text{and} \quad \psi(t) = \alpha(S'(t)).$$

Then $\alpha(S(t)) \le \alpha(T(S)(t))$ and $\alpha(S'(t)) \le \alpha(T(S)'(t))$ for all $t \in I$. As in the proof of Theorem 5.8, we obtain

$$\begin{aligned} \varphi(t) &\le \alpha \left(\int_0^1 G(t, \tau) f(\tau, S(\tau), S'(\tau)) \, d\tau \right) \\ &\le \alpha \left(\int_0^1 G(t, \tau) g(\tau, S(\tau), S'(\tau)) \, d\tau \right) \\ &\le \int_0^1 |G(t, \tau)| w(\tau, \varphi(\tau), \psi(\tau)) \, d\tau. \end{aligned}$$

Similarly,

$$\psi(t) \le \int_0^1 |G_t(t, \tau)| w(\tau, \varphi(\tau), \psi(\tau)) \, d\tau.$$

Thus, φ and ψ satisfy (5.17). Consequently, $\varphi(t) = \psi(t) = 0$. Thus S is relatively compact in C^1. $\square$

Theorem 5.10 yields the following result.

Theorem 5.11 *Let $(E, \langle ., . \rangle)$ be a Hilbert space and $f : I \times E^2 \to E$ continuous. Assume $f = g + h$, where g is uniformly continuous on bounded sets. In addition suppose that $a + b > 0$ and the following conditions hold:*

(i) *for each $R > 0$, there exists a Kamke function w_R, $\eta_R \in L^1(I; \mathbf{R}_+)$ and $\Psi_R \subset E$ compact such that*

$$\alpha(g(t, S)) \leq w_R(t, \alpha(S_1), \alpha(S_2))$$

for all $t \in I$ and any $S = (S_1, S_2) \subset \overline{B}_R(0; E)^2$, and

$$h(t, u, v) \in \eta_R(t) \Psi_R$$

for all $|u| \leq R$, $|v| \leq R$ and a.e. $t \in I$;

(ii) *there exists $r > 0$ such that*

$$\langle u, f(t, u, v) \rangle + |v|^2 > 0 \quad \text{for all } t \in I,$$

whenever $|u| \geq r$ and $\langle u, v \rangle = 0$.

Then there exists $u \in C^2(I; E)$ that solves (5.15).

Remark 5.8 *In Theorem 2.9, $h \equiv 0$ and $w(t, u, v) = \phi(t)(A_0 u + A_1 v)$, where ϕ, A_0 and A_1 satisfy inequality (2.30). To show that this w is a Kamke function, we have only to see that (5.17) implies*

$$\varphi(t) \leq \int_0^1 |G(t, s)| \, \phi(s) \, (A_0 \varphi(s) + A_1 \psi(s)) \, ds$$

$$\leq \left(\int_0^1 |G(t, s)|^q \, \phi(s)^q \, ds \right)^{1/q} \left(A_0 \|\varphi\|_p + A_1 \|\psi\|_p \right)$$

and similarly

$$\psi(t) \leq \int_0^1 |G_t(t, s)| \, \phi(s) \, (A_0 \varphi(s) + A_1 \psi(s)) \, ds$$

$$\leq \left(\int_0^1 |G_t(t, s)|^q \, \phi(s)^q \, ds \right)^{1/q} \left(A_0 \|\varphi\|_p + A_1 \|\psi\|_p \right).$$

It follows that

$$A_0 \|\varphi\|_p + A_1 \|\psi\|_p \leq \rho_p \left(A_0 \|\varphi\|_p + A_1 \|\psi\|_p \right),$$

where ρ_p is given by (2.31). Since $\rho_p < 1$, we have $A_0 \|\varphi\|_p + A_1 \|\psi\|_p = 0$. Hence $\varphi \equiv 0$ and $\psi \equiv 0$.

Leray-Schauder type theorems for other classes of maps involving compactness can be found in [104], [108]-[113], [131]-[132], [136]. For other applications see [3], [14], [16], [21], [26], [37]-[39], [57]-[60], [67]-[69], [79]-[82], [86], [97], [99]-[100], [103], [114]-[115], [117], [128]-[129], [133], [157]. For other methods into the theory of nonlinear differential and integral equations see [10], [12], [27], [64], [74], [78], [87].

6. Applications to Semilinear Elliptic Problems

In this chapter we present applications of the Leray-Schauder type theorems to the weak solvability of the semilinear Dirichlet problem

$$\begin{cases} -\Delta u = cu + f(x, u, \nabla u) & \text{on } \Omega \\ u = 0 & \text{on } \partial\Omega \end{cases} \qquad (6.1)$$

under the assumption that the constant c is not an eigenvalue of $-\Delta$ (*nonresonance* condition) and that the growth of $f(x, u, v)$ on u and v is at most linear. Here Δ is the *Laplacian*

$$\Delta u = \sum_{k=1}^{n} \frac{\partial^2 u}{\partial x_k^2}$$

and ∇ is the gradient

$$\nabla u = \left(\frac{\partial u}{\partial x_1}, \frac{\partial u}{\partial x_2}, ..., \frac{\partial u}{\partial x_n} \right).$$

Such results have been obtained by many authors, see [25], [29], [70], [98] for example. We particularly refer the reader to the papers [62], [91] and [137].

More exactly, we consider Ω a bounded open set of $\mathbf{R}^n$ and $f : \Omega \times \mathbf{R}^{n+1} \to \mathbf{R}$ a function satisfying the *Carathéodory conditions*, i.e.

$f(.,w)$ is measurable for each $w \in \mathbf{R}^{n+1}$ and $f(x,.)$ is continuous for a.e. $x \in \Omega$, and the growth restriction

$$|f(x,u,v)| \leq a|u| + b|v| + h(x) \qquad (6.2)$$

for all $u \in \mathbf{R}$, $v \in \mathbf{R}^n$ and a.e. $x \in \Omega$, where a and b are nonnegative constants and $h \in L^2(\Omega; \mathbf{R}_+)$.

We look for a *weak solution* to (6.1), that is a function $u \in H_0^1(\Omega)$ with

$$\int_\Omega \nabla u \cdot \nabla v \, dx = \int_\Omega (cu + f(x,u,\nabla u)) \, v \, dx \quad \text{for all } v \in H_0^1(\Omega).$$

6.1 Basic Results from the Theory of Linear Elliptic Equations

We shall use basic facts from the L^2 theory of the linear elliptic equations (for details, we refer the reader to [17], [28], [52] and [144]).

a) *Sobolev spaces.* For any integer $m \geq 1$, we denote by $H^m(\Omega)$ the set of all real functions u defined in Ω such that u and its distributional derivatives $D^\alpha u$ of order $|\alpha| \leq m$ all belong to $L^2(\Omega)$. $H^m(\Omega)$ is a Hilbert space with the inner product

$$\langle u, v \rangle_{m,2} = \sum_{|\alpha| \leq m} \langle D^\alpha u, D^\alpha v \rangle_2.$$

Let $C_0^\infty(\Omega)$ denote the space of all functions $u \in C^\infty(\Omega)$ with compact support in Ω. The closure of $C_0^\infty(\Omega)$ in $H^1(\Omega)$ will be denoted by $H_0^1(\Omega)$. Since Ω is supposed to be bounded, we may consider on $H_0^1(\Omega)$ another inner product and the corresponding norm:

$$\langle u, v \rangle_{0,2} = \int_\Omega \nabla u \cdot \nabla v \, dx, \quad \|u\|_{0,2}^2 = \int_\Omega |\nabla u|^2 \, dx.$$

where we have denoted by $\nabla u \cdot \nabla v$ the euclidian inner product of ∇u and ∇v. Poincaré inequality guarantees that $\|.\|_{1,2}$ and $\|.\|_{0,2}$ are equivalent norms on $H_0^1(\Omega)$.

b) *The Rellich-Kondrachov theorem.* 1) The imbedding of $H_0^1(\Omega)$ into $L^2(\Omega)$ is completely continuous. 2) If in addition Ω is C^1, then the imbedding of $H^1(\Omega)$ into $L^2(\Omega)$ is completely continuous.

c) *The inverse of* $-\Delta$. For each $v \in L^2(\Omega)$ there exists a unique $u \in H_0^1(\Omega)$ with

$$\langle u, w \rangle_{0,2} = \langle v, w \rangle_2 \quad \text{for all } w \in H_0^1(\Omega). \tag{6.3}$$

Let $(-\Delta)^{-1} v = u$. The map

$$(-\Delta)^{-1} : L^2(\Omega) \to H_0^1(\Omega)$$

is linear bounded and, by the Rellich-Kondrachov theorem, $(-\Delta)^{-1}$ is completely continuous from $L^2(\Omega)$ into $L^2(\Omega)$.

d) *Regularity of weak solutions.* If Ω is C^2, then $(-\Delta)^{-1}(L^2(\Omega)) \subset H^2(\Omega)$ and the linear map $(-\Delta)^{-1}$ is also bounded from $L^2(\Omega)$ into $H^2(\Omega)$.

e) *Eigenvalues.* Let

$$0 < \lambda_1 < \lambda_2 \leq \lambda_3 \leq \ldots \leq \lambda_k \leq \ldots$$

be the sequence of all eigenvalues of $-\Delta$. We know that

$$1/\lambda_1 = \sup\left\{\left\|(-\Delta)^{-1} v\right\|_2; \ v \in L^2(\Omega), \|v\|_2 = 1\right\}.$$

Also note that

$$\lambda_1 = \inf\left\{\|u\|_{0,2}^2 / \|u\|_2^2; \ u \in H_0^1(\Omega) \setminus \{0\}\right\}. \tag{6.4}$$

f) *Eigenfunctions.* There exists a *Hilbert base* (orthonormal and complete) $(\varphi_k)_{k \geq 1}$ in $L^2(\Omega)$ of eigenvalues. Hence

$$\langle \varphi_k, \varphi_j \rangle_2 = \begin{cases} 0 \text{ for } k \neq j \\ 1 \text{ for } k = j, \end{cases}$$

$$\langle \varphi_k, w \rangle_{0,2} = \lambda_k \langle \varphi_k, w \rangle_2 \quad \text{for all } w \in H_0^1(\Omega)$$

and

$$v = \sum_{k=1}^{\infty} \langle v, \varphi_k \rangle_2 \, \varphi_k \quad \text{for all } v \in L^2(\Omega)$$

Moreover, the sequence $\left(\lambda_k^{-1/2} \varphi_k\right)_{k \geq 1}$ is a Hilbert base in $(H_0^1(\Omega), \langle ., . \rangle_{0,2})$. Thus, for each $u \in H_0^1(\Omega)$, one has

$$u = \sum_{k=1}^{\infty} c_k \varphi_k, \text{ where } c_k = \langle u, \varphi_k \rangle_{0,2} / \lambda_k = \langle u, \varphi_k \rangle_2. \tag{6.5}$$

Notice that for each k, φ_k is analytic on Ω and consequently, $\varphi_k^2(x) > 0$ for a.e. $x \in \Omega$. Also recall that $\varphi_1(x) > 0$ for all $x \in \Omega$.

g) *The Nemitskii superposition operator.* If $g : \Omega \times \mathbf{R}^m \to \mathbf{R}$ satisfies the Carathéodory conditions and the growth condition

$$|g(x, u)| \leq C |u| + h(x)$$

for all $u \in \mathbf{R}^m$ and a.e. $x \in \Omega$, where C is a nonnegative constant and $h \in L^2(\Omega; \mathbf{R}_+)$, then the Nemitskii operator

$$u \longmapsto g(., u(.))$$

is well defined, bounded (sends bounded sets into bounded sets) and continuous from $L^2(\Omega; \mathbf{R}^m)$ into $L^2(\Omega)$.

h) *An auxiliary result.*

Lemma 6.1 *Let c be any constant with $c \neq \lambda_k$ for $k = 1, 2, \ldots$. For each $v \in L^2(\Omega)$, there exists a unique weak solution $u \in H_0^1(\Omega)$ to the problem*

$$\begin{cases} Lu := -\Delta u - cu = v & \text{on } \Omega \\ u = 0 & \text{on } \partial\Omega \end{cases}$$

denoted by $L^{-1}v$, and the following eigenfunction expansion holds

$$L^{-1}v = \sum_{k=1}^{\infty} (\lambda_k - c)^{-1} \langle v, \varphi_k \rangle_2 \varphi_k \tag{6.6}$$

where the series converges in $H_0^1(\Omega)$. In addition,

$$\left\| L^{-1}v \right\|_2 \leq \mu_c \|v\|_2 \quad \text{for all } v \in L^2(\Omega) \tag{6.7}$$

where

$$\mu_c = \max \left\{ |\lambda_k - c|^{-1} ; k = 1, 2, \ldots \right\}.$$

Proof. We first prove the convergence of the series (6.6). Since $\left(\lambda_k^{-1/2} \varphi_k \right)_{k \geq 1}$ is a Hilbert base in $\left(H_0^1(\Omega), \langle ., . \rangle_{0,2} \right)$, we have

$$\left\| \sum_{k=m+1}^{m+p} (\lambda_k - c)^{-1} \langle v, \varphi_k \rangle_2 \varphi_k \right\|_{0,2}^2$$

$$= \sum_{k=m+1}^{m+p} \langle v, \varphi_k \rangle_2^2 \lambda_k / (\lambda_k - c)^2 \le C \sum_{k=m+1}^{m+p} \langle v, \varphi_k \rangle_2^2$$

where C is a constant such that $\lambda_k / (\lambda_k - c)^2 \le C$ for all k. Thus the convergence of (6.6) follows from the convergence of the numerical series $\sum \langle v, \varphi_k \rangle_2^2$ (Bessel's inequality). Let $u \in H_0^1 (\Omega)$ be the sum of series (6.6). Next we check that $Lu = v$ weakly, i.e.

$$\langle u, w \rangle_{0,2} - c \langle u, w \rangle_2 = \langle v, w \rangle_2 \quad \text{for all } w \in H_0^1 (\Omega).$$

Indeed, we have

$$\langle u, w \rangle_{0,2} = \sum_{k=1}^{\infty} (\lambda_k - c)^{-1} \langle v, \varphi_k \rangle_2 \langle \varphi_k, w \rangle_{0,2}$$

$$= \sum_{k=1}^{\infty} \lambda_k (\lambda_k - c)^{-1} \langle v, \varphi_k \rangle_2 \langle \varphi_k, w \rangle_2$$

and

$$\langle u, w \rangle_2 = \sum_{k=1}^{\infty} (\lambda_k - c)^{-1} \langle v, \varphi_k \rangle_2 \langle \varphi_k, w \rangle_2 .$$

Hence

$$\langle u, w \rangle_{0,2} - c \langle u, w \rangle_2 = \sum_{k=1}^{\infty} \langle v, \varphi_k \rangle_2 \langle \varphi_k, w \rangle_2$$

$$= \left\langle \sum_{k=1}^{\infty} \langle v, \varphi_k \rangle_2 \varphi_k, w \right\rangle_2 = \langle v, w \rangle_2$$

as desired.

The uniqueness follows from $c \ne \lambda_k$, $k = 1, 2, \ldots$.

To prove (6.7), observe that

$$\left\| \sum_{k=1}^{m} (\lambda_k - c)^{-1} \langle v, \varphi_k \rangle_2 \varphi_k \right\|_2^2 \to \left\| L^{-1} v \right\|_2^2 \quad \text{as } m \to \infty$$

and, on the other hand,

$$\left\| \sum_{k=1}^{m} (\lambda_k - c)^{-1} \langle v, \varphi_k \rangle_2 \varphi_k \right\|_2^2 = \sum_{k=1}^{m} (\lambda_k - c)^{-2} \langle v, \varphi_k \rangle_2^2$$

$$\le \mu_c^2 \sum_{k=1}^{m} \langle v, \varphi_k \rangle_2^2 \to \mu_c^2 \|v\|_2^2 .$$

$\square$

6.2 Applications of the Banach, Schauder and Darbo Fixed Point Theorems

Fixed point formulations of problem (6.1)

1) We can write (6.1) as the following fixed point problem on $H_0^1(\Omega)$:
$A(u) = u$, where
$$A : H_0^1(\Omega) \to H_0^1(\Omega), \quad A = L^{-1}F;$$

here
$$F : H_0^1(\Omega) \to L^2(\Omega), \quad F(u)(x) = f(x, u(x), \nabla u(x)).$$

Clearly, any fixed point of A belongs to the subspace $(-\Delta)^{-1}(L^2(\Omega))$ of $H_0^1(\Omega)$.

2) If we look a priori for a solution u of the form $u = L^{-1}v$ with $v \in L^2(\Omega)$, hence in the subspace $(-\Delta)^{-1}(L^2(\Omega))$, then we have to solve a fixed point problem on $L^2(\Omega) : T(v) = v$, where
$$T : L^2(\Omega) \to L^2(\Omega), \quad T(v) = f(., L^{-1}v, \nabla L^{-1}v). \tag{6.8}$$

We now show how the fixed point theorems of Banach, Schauder and Darbo can be used to obtain existence results for problem (6.1). Also we shall see that better results can be obtained if we use Leray-Schauder type theorems.

Theorem 6.2 *Suppose*
$$\lambda_j < c < \lambda_{j+1} \quad \text{for some } j \in \mathbf{N}, \ j \geq 1, \ \text{or} \ \ 0 \leq c < \lambda_1. \tag{6.9}$$

Also assume that f satisfies the Carathéodory conditions, $f(., 0, 0) \in L^2(\Omega)$ and
$$|f(x, u, v) - f(x, \bar{u}, \bar{v})| \leq a|u - \bar{u}| + b|v - \bar{v}| \tag{6.10}$$

for all u, $\bar{u} \in \mathbf{R}$, v, $\bar{v} \in \mathbf{R}^n$ and a.e. $x \in \Omega$, where a, b are two nonnegative constants with
$$a\mu_c + b\sqrt{\mu_c(1 + c\mu_c)} < 1. \tag{6.11}$$

Then (6.1) has a unique solution $u \in H_0^1(\Omega)$. In addition
$$T^k(w) \to v \text{ in } L^2(\Omega) \text{ as } k \to \infty$$

for all $w \in L^2(\Omega)$, where $u = L^{-1}v$.

Proof. Using g) we immediately see that the map (6.8) is well defined. Let us show that T is in fact a contraction. For this, let v_1, $v_2 \in L^2(\Omega)$. Then, from (6.10), we have

$$\|T(v_1) - T(v_2)\|_2 \leq a \left\|L^{-1}(v_1 - v_2)\right\|_2 + b \left\|L^{-1}(v_1 - v_2)\right\|_{0,2}.$$

From (6.7), we have

$$\left\|L^{-1}(v_1 - v_2)\right\|_2 \leq \mu_c \|v_1 - v_2\|_2$$

which together with (6.3), yields

$$\left\|L^{-1}(v_1 - v_2)\right\|_{0,2}^2 = c\left\|L^{-1}(v_1 - v_2)\right\|_2^2$$

$$+ \left\langle v_1 - v_2, L^{-1}(v_1 - v_2)\right\rangle_2 \leq c\mu_c^2 \|v_1 - v_2\|_2^2 + \mu_c \|v_1 - v_2\|_2^2.$$

Thus

$$\left\|L^{-1}(v_1 - v_2)\right\|_{0,2} \leq \sqrt{\mu_c(1 + c\mu_c)} \|v_1 - v_2\|_2. \qquad (6.12)$$

Consequently

$$\|T(v_1) - T(v_2)\|_2 \leq \left(a\mu_c + b\sqrt{\mu_c(1 + c\mu_c)}\right) \|v_1 - v_2\|_2.$$

This together with (6.11), shows that T is a contraction. The conclusion follows from Banach's fixed point theorem. $\square$

Remark 6.1 *If $c = 0$, we have $\mu_c = 1/\lambda_1$ and (6.11) reduces to the inequality*

$$\frac{a}{\lambda_1} + \frac{b}{\sqrt{\lambda_1}} < 1.$$

This case was discussed in [[62], Theorem 1].

Notice in Theorem 6.2, that no smoothness assumption on Ω is required.

Theorem 6.3 *Suppose that Ω is C^2, (6.9) holds and f satisfies the Carathéodory conditions and (6.2) for a, b as in (6.11). Then (6.1) has at least one solution $u \in H^2(\Omega) \cap H_0^1(\Omega)$.*

Proof. By d), $\nabla L^{-1}v \in H^1(\Omega; \mathbf{R}^n)$ and so,

$$\left(L^{-1}v, \nabla L^{-1}v\right) \in H^1\left(\Omega; \mathbf{R}^{n+1}\right).$$

Next, by the Rellich-Kondrachov theorem, the imbedding of $H^1(\Omega; \mathbf{R}^{n+1})$ into $L^2(\Omega; \mathbf{R}^{n+1})$ is completely continuous and since the Nemitskii operator $f(., u(.))$ from $L^2(\Omega; \mathbf{R}^{n+1})$ into $L^2(\Omega)$ is continuous and bounded, it follows that T is completely continuous from $L^2(\Omega)$ into itself. On the other hand, similar estimates to those in the proof of Theorem 6.2 show that

$$\|T(v)\|_2 \leq a\left\|L^{-1}v\right\|_2 + b\left\|L^{-1}v\right\|_{0,2} + \|h\|_2$$

$$\leq \left(a\mu_c + b\sqrt{\mu_c(1+c\mu_c)}\right)\|v\|_2 + \|h\|_2.$$

Now (6.11) guarantees that T is a self-map of a sufficiently large closed ball of $L^2(\Omega)$. Thus we may apply Schauder's fixed point theorem. $\square$

If $c = 0$, Theorem 6.3 reduces to a result in [[62], Theorem 2].

The smoothness of Ω was required to show the complete continuity of T, in the case when f depends on ∇u. In the next theorem the map T will be only a set-contraction and the smoothness assumption on Ω will not be necessary.

Theorem 6.4 *Suppose (6.9) holds and that f has the decomposition*

$$f(x, u, v) = f_0(x, u, v) + f_1(x, u),$$

where f_0 and f_1 satisfy the Carathéodory conditions, $f_0(., 0, 0) \in L^2(\Omega)$ and the following inequalities are true

$$|f_0(x, u, v) - f_0(x, \bar{u}, \bar{v})| \leq a_0|u - \bar{u}| + b|v - \bar{v}|$$

$$|f_1(x, u)| \leq a_1|u| + h(x)$$

for all $u, \bar{u} \in \mathbf{R}$, $v, \bar{v} \in \mathbf{R}^n$ and a.e. $x \in \Omega$, where $h \in L^2(\Omega)$. If $a = a_0 + a_1$ and b satisfy (6.11), then there exists $u \in H_0^1(\Omega)$ that solves (6.1).

Proof. We decompose T as follows $T = T_0 + T_1$,

$$T_0(v) = f_0(., L^{-1}v, \nabla L^{-1}v), \quad T_1(v) = f_1(., L^{-1}v)$$

where T_1 is completely continuous and T_0 is a contraction. Hence T is a set-contraction. From (6.11), T maps a sufficiently large ball of $L^2(\Omega)$ into itself. Thus we may apply Darbo's fixed point theorem. □

For example, if $f = f(u)$, where $f \in C(\mathbf{R})$ and $f(u)/u \to 0$ as $|u| \to \infty$, then we may choose $a_0 = b = 0$ and $a = a_1$ small enough so that $a\mu_c < 1$. Thus, (6.11) is automatically satisfied.

6.3 Applications of the Leray-Schauder Type Theorems

Theorem 6.5 *Suppose that Ω is C^2 and that f has the decomposition*

$$f(x, u, v) = g(x, u, v) u + f_0(x, u, v) + f_1(x, u, v)$$

where g, f_0, f_1 satisfy the Carathéodory conditions. Also assume that:

$$|f_0(x, u, v)| \leq a|u| + b|v| + h_0(x) \tag{6.13}$$

$$|f_1(x, u, v)| \leq a_1|u| + b_1|v| + h_1(x) \tag{6.14}$$

$$uf_1(x, u, v) \leq 0 \tag{6.15}$$

$$-M \leq g(x, u, v) + c \leq \beta < \lambda_1 \tag{6.16}$$

for all $u \in \mathbf{R}$, $v \in \mathbf{R}^n$ and a.e. $x \in \Omega$, where a, b, a_1, b_1, β, $M \in \mathbf{R}_+$ and h_0, $h_1 \in L^2(\Omega; \mathbf{R}_+)$. In addition assume that $0 \leq c \leq \beta$ and

$$a/\lambda_1 + b/\sqrt{\lambda_1} < 1 - \beta/\lambda_1. \tag{6.17}$$

Then (6.1) has at least one solution $u \in H^2(\Omega) \cap H_0^1(\Omega)$.

Proof. We look for a fixed point $v \in L^2(\Omega)$ of T. As above, $T : L^2(\Omega) \to L^2(\Omega)$ is a completely continuous map. Now we show that the set of all solutions to

$$v = \lambda T(v) \tag{6.18}$$

when $\lambda \in [0, 1]$, is bounded in $L^2(\Omega)$. Let $v \in L^2(\Omega)$ be any solution to (6.18). Let $u = L^{-1}v$. It is clear that u solves

$$\begin{cases} -\Delta u - cu = \lambda f(x, u, \nabla u) & \text{on } \Omega \\ u = 0 & \text{on } \partial\Omega. \end{cases} \tag{6.19}$$

Since u is a weak solution of (6.19), we have

$$\|u\|_{0,2}^2 = \langle cu + \lambda f(x, u, \nabla u), u \rangle_2.$$

Using (6.15), (6.16) and $c \leq \beta$, we obtain

$$R(u) := \|u\|_{0,2}^2 - \beta \|u\|_2^2 \tag{6.20}$$

$$\leq \|u\|_{0,2}^2 - \langle cu + \lambda g(x, u, \nabla u) u, u \rangle_2 \leq |\langle f_0(x, u, \nabla u), u \rangle_2|.$$

From (6.5), we see that

$$R(u) = \sum_{k=1}^{\infty} (\lambda_k - \beta) c_k^2 \geq \sum_{k=1}^{\infty} \lambda_k (1 - \beta/\lambda_1) c_k^2 \tag{6.21}$$

$$= (1 - \beta/\lambda_1) \|u\|_{0,2}^2.$$

Return to (6.20) and use (6.4), (6.13) and Hölder's inequality to obtain

$$(1 - \beta/\lambda_1) \|u\|_{0,2}^2 \leq \int_{\Omega} \left(au^2 + b|\nabla u| \, |u| + h_0 \, |u| \right) dx$$

$$\leq \left(a/\lambda_1 + b/\sqrt{\lambda_1} \right) \|u\|_{0,2}^2 + C \|u\|_{0,2}$$

for some constant $C > 0$. Thus (6.17) guarantees that there is a constant $r > 0$ independent of λ with $\|u\|_{0,2} \leq r$. Finally, a bound for $\|v\|_2$ can be immediately derived from (6.18). The conclusion now follows from Theorem 5.3. $\square$

Remark 6.2 *If in Theorem 6.5 f does not depend on ∇u, then the existence of at least one solution $u \in H_0^1(\Omega)$ to (6.1) follows without any assumption on the smoothness of Ω.*

When $g = f_1 = 0$, Theorem 6.5 reduces to Theorem 6.3 for $j = 0$. Indeed, we have $\beta = c < \lambda_1$, $\mu_c = 1/(\lambda_1 - c)$ and it is easily checked that (6.17) is equivalent to (6.11).

For $g \equiv 0$, $c = 0$ and $f_0 = h_0$, Theorem 6.5 was established in [[137], Theorem 4].

If only f_0 depends on ∇u and f_0 satisfies a Lipschitz condition, we can prove a similar existence result by means of the Leray-Schauder theorem for set-contractions without the smoothness hypothesis on Ω.

Theorem 6.6 *Suppose that f has the decomposition*

$$f(x, u, v) = g(x, u) u + f_0(x, u, v) + f_1(x, u)$$

where g, f_0, f_1 satisfy the Carathéodory conditions and $f_0(., 0, 0) \in L^2(\Omega)$. Also assume that the following conditions are satisfied:

$$|f_0(x, u, v) - f_0(x, \bar{u}, \bar{v})| \leq a |u - \bar{u}| + b |v - \bar{v}|$$

$$|f_1(x, u)| \leq a_1 |u| + h(x)$$

$$u f_1(x, u) \leq 0$$

$$-M \leq g(x, u) + c \leq \beta < \lambda_1$$

for all u, $\bar{u} \in \mathbf{R}$, v, $\bar{v} \in \mathbf{R}^n$ and a.e. $x \in \Omega$, where a, b, a_1, β, $M \in \mathbf{R}_+$ and $h \in L^2(\Omega; \mathbf{R}_+)$. In addition assume that $0 \leq c \leq \beta$ and (6.17) holds. Then (6.1) has at least one solution $u \in H_0^1(\Omega)$.

Proof. Let $T = T_0 + T_1$, where

$$T_0(v) = f_0\left(., L^{-1}v, \nabla L^{-1}v\right)$$

and

$$T_1(v) = g\left(., L^{-1}v\right) L^{-1}v + f_1\left(., L^{-1}v\right)$$

for $v \in L^2(\Omega)$. Then T_1 is a completely continuous map, while T_0 is a contraction since (6.17) implies (6.11). Hence T is a set-contraction on $L^2(\Omega)$. Next the a priori bound of solutions is obtained by essentially the same reasoning as in Theorem 6.5. $\square$

Notice that when $g = f_1 = 0$, Theorem 6.6 reduces to Theorem 6.2 for $0 \leq c < \lambda_1$.

Theorem 6.7 *Suppose that Ω is C^2 and f has the decomposition*

$$f(x, u, v) = g(x, u, v) u + f_0(x, u, v) + f_1(x, u, v),$$

where g, f_0 and f_1 satisfy the Carathéodory conditions. Also assume that conditions (6.14), (6.15) and

$$|f_0(x, u, v)| \leq a |u|^\gamma + b |v|^\eta + h_0(x) \tag{6.22}$$

$$-M \leq g(x, u, v) + c \leq \beta(x) \leq \lambda_1 \tag{6.23}$$

are satisfied for all $u \in \mathbf{R}$, $v \in \mathbf{R}^n$ and a.e. $x \in \Omega$, where a, b, a_1, b_1, $M \in \mathbf{R}_+$, $\beta \in L^\infty(\Omega)$, h_0, $h_1 \in L^2(\Omega; \mathbf{R}_+)$ and γ, $\eta \in [0,1)$. In addition assume that $0 \leq c \leq \beta(x)$ a.e. on Ω and $\beta(x) < \lambda_1$ on a subset of Ω of positive measure. Then problem (6.1) has at least one solution $u \in H^2(\Omega) \cap H_0^1(\Omega)$.

Proof. As above, $T : L^2(\Omega) \to L^2(\Omega)$ is a completely continuous map. Now we show that the set of all solutions to (6.18) is bounded in $L^2(\Omega)$. Let $v \in L^2(\Omega)$ be any solution to (6.18) and $u = L^{-1}v$. Since u is a weak solution of (6.19), we have

$$\|u\|_{0,2}^2 = \langle cu + \lambda f(x, u, \nabla u), u \rangle_2 .$$

Using (6.15) and (6.23), we obtain

$$R(u) := \|u\|_{0,2}^2 - \int_\Omega \beta(x) u^2 dx \leq |\langle f_0(x, u, \nabla u), u \rangle_2| . \qquad (6.24)$$

Now we show that there exists $\epsilon > 0$ with

$$R(u) \geq \epsilon \|u\|_{0,2}^2 .$$

First we note that $R(u) \geq 0$ for all $u \in H_0^1(\Omega)$ and $R(u) = 0$ if and only if $u = 0$. Indeed, using (6.5) and (6.23), we have

$$R(u) \geq \sum_{k=1}^\infty (\lambda_k - \lambda_1) c_k^2 \geq 0.$$

If $R(u) = 0$, then the above inequalities imply $c_k = 0$ for all $k \geq 2$, that is $u = c_1 \varphi_1$. So,

$$0 = R(u) = c_1^2 \int_\Omega \left(|\nabla \varphi_1|^2 - \beta(x) \varphi_1^2 \right) dx$$

$$= c_1^2 \int_\Omega (\lambda_1 - \beta(x)) \varphi_1^2 dx.$$

Since $\varphi_1(x) > 0$ on Ω and $\lambda_1 - \beta(x) > 0$ on a set of positive measure, this implies that $c_1 = 0$ too. Hence $u = 0$.

Suppose now that such an $\epsilon > 0$ does not exist. Then, there is a sequence (u_k) with

$$\|u_k\|_{0,2} = 1 \quad \text{and} \quad R(u_k) \to 0 \text{ as } k \to \infty.$$

Due to the compact imbedding of $H_0^1(\Omega)$ into $L^2(\Omega)$, we may suppose, passing if necessary to a subsequence, that

$$u_k \to u \text{ in } L^2(\Omega) \text{ and } u_k \rightharpoonup u \text{ (weakly) in } H_0^1(\Omega).$$

It is known that $\|u\|_{0,2} \leq \liminf \|u_k\|_{0,2}$. Letting $k \to \infty$, we obtain $R(u) \leq 0$. Hence $u = 0$. Then

$$\int_\Omega \beta(x) u_k^2 dx \to 0 \quad \text{as } k \to \infty.$$

This yields

$$1 = \|u_k\|_{0,2}^2 = R(u_k) + \int_\Omega \beta(x) u_k^2 dx \to 0 \quad \text{as } k \to \infty,$$

a contradiction.

Returning to (6.24) and using (6.22) and Hölder's inequality, we easily obtain

$$\epsilon \|u\|_{0,2}^2 \leq C_1 \|u\|_{0,2}^\theta + C_2 \|u\|_{0,2}$$

for some constants C_1 and C_2; here $\theta = \max\{\gamma + 1, \eta + 1\} < 2$. Thus, there is a constant $r > 0$ independent of λ with $\|u\|_{0,2} \leq r$. Finally, a bound for $\|v\|_2$ can be immediately derived from (6.18). The conclusion now follows from Theorem 5.3. $\square$

Remark 6.3 *If in Theorem 6.7 f does not depend on ∇u, then the existence of at least one solution $u \in H_0^1(\Omega)$ to (6.1) follows without any assumption on the smoothness of Ω.*

For the remainder of this section we consider the case when $g + c$ lies between two eigenvalues of $-\Delta$.

Theorem 6.8 *Suppose that Ω is C^2 and f has the decomposition*

$$f(x, u, v) = g(x, u, v)u + f_0(x, u, v)$$

where g and f_0 satisfy the Carathéodory conditions. Also assume that the following conditions are satisfied:

$$|f_0(x, u, v)| \leq a|u| + b|v| + h(x) \tag{6.25}$$

$$\lambda_j < \beta_1 \leq g(x, u, v) + c \leq \beta_2 < \lambda_{j+1} \tag{6.26}$$

for all $u \in \mathbf{R}$, $v \in \mathbf{R}^n$ and a.e. $x \in \Omega$, where a, b, β_1, $\beta_2 \in \mathbf{R}_+$, $h \in L^2(\Omega; \mathbf{R}_+)$, $j \geq 1$ and $\beta_1 \leq c \leq \beta_2$. If

$$a/\lambda_1 + b/\sqrt{\lambda_1} < \min\{\beta_1/\lambda_j - 1, 1 - \beta_2/\lambda_{j+1}\}, \tag{6.27}$$

then problem (6.1) has at least one solution $u \in H^2(\Omega) \cap H_0^1(\Omega)$.

Proof. Let $v \in L^2(\Omega)$ be any solution to (6.18) and $u = L^{-1}v$. Since $\left(\lambda_k^{-1/2}\varphi_k\right)_{k \geq 1}$ is a Hilbert base for $H_0^1(\Omega)$, we may decompose $H_0^1(\Omega)$ as follows:

$$H_0^1(\Omega) = X_1 \oplus X_2,$$

where X_1 is the subspace generated by the first j eigenfunctions φ_1, φ_2, ..., φ_j and $X_2 = X_1^\perp$. Let $u = y + z$ with $y \in X_1$ and $z \in X_2$. Then

$$y = \sum_{k=1}^{j} c_k \varphi_k, \quad z = \sum_{k=j+1}^{\infty} c_k \varphi_k$$

where $c_k = \langle u, \varphi_k \rangle_{0,2} / \lambda_k$.

Since u is a weak solution to (6.19), we have

$$\langle u, z - y \rangle_{0,2} = \langle cu + \lambda f(x, u, \nabla u), z - y \rangle_2$$

and so

$$\|z\|_{0,2}^2 - c\|z\|_2^2 - \lambda \langle g(x, u, \nabla u) z, z \rangle_2$$
$$- \|y\|_{0,2}^2 + c\|y\|_2^2 + \lambda \langle g(x, u, \nabla u) y, y \rangle_2$$
$$= \lambda \langle f_0(x, u, \nabla u), z - y \rangle_2.$$

Now, from (6.26),

$$c + \lambda g = \lambda(c + g) + (1 - \lambda)c \leq \lambda\beta_2 + (1 - \lambda)\beta_2 = \beta_2$$

and so

$$-c\|z\|_2^2 - \lambda \langle g(x, u, \nabla u) z, z \rangle_2 \geq -\beta_2 \|z\|_2^2$$

Similarly

$$c\|y\|_2^2 + \lambda \langle g(x, u, \nabla u) y, y \rangle_2 \geq \beta_1 \|y\|_2^2.$$

It follows that

$$\|z\|_{0,2}^2 - \beta_2 \|z\|_2^2 - \|y\|_{0,2}^2 + \beta_1 \|y\|_2^2 \tag{6.28}$$

$$\leq |\langle f_0\,(x,u,\nabla u)\,,\,z-y\rangle_2|\,.$$

Let

$$R\,(u) := \|z\|_{0,2}^2 - \beta_2\,\|z\|_2^2 - \|y\|_{0,2}^2 + \beta_1\,\|y\|_2^2\,.$$

Using (6.5), we find that

$$R\,(u) \;=\; \sum_{k=j+1}^{\infty} (\lambda_k - \beta_2)\,c_k^2 + \sum_{k=1}^{j} (\beta_1 - \lambda_k)\,c_k^2$$

$$=\; \sum_{k=j+1}^{\infty} (1 - \beta_2/\lambda_k)\,\lambda_k c_k^2 + \sum_{k=1}^{j} (\beta_1/\lambda_k - 1)\,\lambda_k c_k^2$$

$$\geq\; \min\{\beta_1/\lambda_j - 1,\, 1 - \beta_2/\lambda_{j+1}\}\,\|u\|_{0,2}^2\,.$$

On the other hand, from (6.25),

$$|\langle f_0\,(x,u,\nabla u)\,,\,z-y\rangle_2| \;\leq\; \Big(a\,\|u\|_2 + b\,\|u\|_{0,2} + \|h\|_2\Big)\,\|z-y\|_2$$

and since $\|z - y\|_2 = \|z + y\|_2 = \|u\|_2\,,$ this yields

$$|\langle f_0\,(x,u,\nabla u)\,,\,z-y\rangle_2| \;\leq\; \Big(a/\lambda_1 + b/\sqrt{\lambda_1}\Big)\,\|u\|_{0,2}^2 + C\,\|u\|_{0,2}$$

for some constant $C > 0$. Thus, (6.28) implies that

$$\min\{\beta_1/\lambda_j - 1,\, 1 - \beta_2/\lambda_{j+1}\}\,\|u\|_{0,2}^2$$

$$\leq\; \Big(a/\lambda_1 + b/\sqrt{\lambda_1}\Big)\,\|u\|_{0,2}^2 + C\,\|u\|_{0,2}\,.$$

This with (6.27) guarantees that there exists $r > 0$ independent of λ with $\|u\|_{0,2} \leq r$. Next, as usual, we obtain a bound of $\|v\|_2$ and we apply Theorem 5.3. $\square$

Remark 6.4 *If in Theorem 6.8 f does not depend on ∇u, then the existence of at least one solution $u \in H_0^1\,(\Omega)$ to (6.1) follows without assuming that Ω is C^2.*

The analog of Theorem 6.6 is the following one.

Theorem 6.9 *Suppose that f has the decomposition*

$$f(x, u, v) = g(x, u) u + f_0(x, u, v)$$

where g, f_0 satisfy the Carathéodory conditions and $f_0(\cdot, 0, 0) \in L^2(\Omega)$. Also assume that the following conditions are satisfied:

$$|f_0(x, u, v) - f_0(x, \bar{u}, \bar{v})| \le a |u - \bar{u}| + b |v - \bar{v}|$$

$$\lambda_j < \beta_1 \le g(x, u) + c \le \beta_2 < \lambda_{j+1}$$

for all u, $\bar{u} \in \mathbf{R}$, v, $\bar{v} \in \mathbf{R}^n$ and a.e. $x \in \Omega$, where a, b, β_1, $\beta_2 \in \mathbf{R}_+$, $j \ge 1$ and $\beta_1 \le c \le \beta_2$. In addition assume that (6.11) and (6.27) hold. Then problem (6.1) has at least one solution $u \in H_0^1(\Omega)$.

Proof. Condition (6.11) implies that T is a set-contraction, while (6.27) assures the a priori boundedness of the solutions. $\square$

Finally, let us state the analog of Theorem 6.7.

Theorem 6.10 *Suppose that Ω is C^2 and f has the decomposition*

$$f(x, u, v) = g(x, u, v) u + f_0(x, u, v)$$

where g, f_0 satisfy the Carathéodory conditions. Also assume that (6.22) and

$$\lambda_j \le \beta_1(x) \le g(x, u, v) + c \le \beta_2(x) \le \lambda_{j+1} \qquad (6.29)$$

hold for all $u \in \mathbf{R}$, $v \in \mathbf{R}^n$ and a.e. $x \in \Omega$, where a, $b \in \mathbf{R}_+$, $h_0 \in L^2(\Omega; \mathbf{R}_+)$, γ, $\eta \in [0, 1)$, $j \ge 1$ and β_1, $\beta_2 \in L^\infty(\Omega)$. In addition assume that $\beta_1(x) \le c \le \beta_2(x)$ a.e. on Ω and that $\lambda_j < \beta_1(x)$, $\beta_2(x) < \lambda_{j+1}$ on some subsets of Ω of positive measure. Then (6.1) has at least one solution $u \in H^2(\Omega) \cap H_0^1(\Omega)$.

Proof. As in the proof of Theorem 6.8, we decompose $H_0^1(\Omega)$ as follows: $H_0^1(\Omega) = X_1 \oplus X_2$, where X_1 is the subspace generated by the first j eigenfunctions φ_1, φ_2, ..., φ_j, $X_2 = X_1^\perp$ and we represent any element u as $u = y + z$ with $y \in X_1$ and $z \in X_2$.

Let v be any solution to (6.18) and $u = L^{-1}v$. Since u is a weak solution to (6.19), we have

$$\langle u, z - y \rangle_{0,2} = \langle cu + \lambda f(x, u, \nabla u), z - y \rangle_2$$

and so

$$\|z\|_{0,2}^2 - c\|z\|_2^2 - \lambda \langle g(x, u, \nabla u) z, z \rangle_2$$

$$- \|y\|_{0,2}^2 + c\|y\|_2^2 + \lambda \langle g(x, u, \nabla u) y, y \rangle_2$$

$$= \lambda \langle f_0(x, u, \nabla u), z - y \rangle_2.$$

Now, from (6.29) and $\beta_1(x) \le c \le \beta_2(x)$, we have

$$-c\|z\|_2^2 - \lambda \langle g(x, u, \nabla u) z, z \rangle_2 \ge -\int_\Omega \beta_2(x) z^2 dx$$

and

$$c\|y\|_2^2 + \lambda \langle g(x, u, \nabla u) y, y \rangle_2 \ge \int_\Omega \beta_1(x) y^2 dx.$$

It follows that

$$\|z\|_{0,2}^2 - \int_\Omega \beta_2(x) z^2 dx - \|y\|_{0,2}^2 + \int_\Omega \beta_1(x) y^2 dx \qquad (6.30)$$

$$\le |\langle f_0(x, u, \nabla u), z - y \rangle_2|.$$

Let

$$R(u) := \|z\|_{0,2}^2 - \int_\Omega \beta_2(x) z^2 dx - \|y\|_{0,2}^2 + \int_\Omega \beta_1(x) y^2 dx.$$

We show that there exists $\epsilon > 0$ independent of u, with

$$R(u) \ge \epsilon \|u\|_{0,2}^2. \qquad (6.31)$$

We first note that

$$R(u) \ge \|z\|_{0,2}^2 - \lambda_{j+1} \|z\|_2^2 - \|y\|_{0,2}^2 + \lambda_j \|y\|_2^2$$

$$= \sum_{k=j+1}^\infty (\lambda_k - \lambda_{j+1}) c_k^2 - \sum_{k=1}^j (\lambda_k - \lambda_j) c_k^2 \ge 0.$$

Also, if $R(u) = 0$, then from the above formula, it follows that $c_k = 0$ for all k, $k \neq j$, $k \neq j+1$. Thus, $u = c_j \varphi_j + c_{j+1} \varphi_{j+1}$ and consequently, $y = c_j \varphi_j$ and $z = c_{j+1} \varphi_{j+1}$. Then

$$0 = R(u) = c_{j+1}^2 \int_\Omega (\lambda_{j+1} - \beta_2(x)) \varphi_{j+1}^2 dx$$

$$+ c_j^2 \int_\Omega \left(\beta_1 (x) - \lambda_j \right) \varphi_j^2 dx.$$

Since $\lambda_{j+1} - \beta_2 (x) > 0$ and $\beta_1 (x) - \lambda_j > 0$ on some sets of positive measure and φ_k is analytic on Ω, we have that $c_j = c_{j+1} = 0$. Thus, $u = 0$.

Now suppose (6.31) is false for all $\epsilon > 0$. Then there exists a sequence (u_k) with

$$\|u_k\|_{0,2} = 1 \quad \text{and} \quad R(u_k) \to 0 \text{ as } k \to \infty.$$

Let $u_k = y_k + z_k$, where $y_k \in X_1$ and $z_k \in X_2$. Passing to a subsequence, we may assume that $u_k \to u$ in $L^2 (\Omega)$ and $u_k \rightharpoonup u$ in $H_0^1 (\Omega)$. Also, if $u = y + z$, then $y_k \to y$ in $H_0^1 (\Omega)$, $z_k \to z$ in $L^2 (\Omega)$ and $z_k \rightharpoonup z$ in $H_0^1 (\Omega)$. Note that $\|z\|_{0,2} \leq \liminf \|z_k\|_{0,2}$. Then, it follows that $R(u) \leq 0$. Hence $u = 0$ and in consequence, $y = z = 0$. However,

$$1 = \|u_k\|_{0,2}^2 = R(u_k) + 2 \|y_k\|_{0,2}^2$$

$$+ \int_\Omega \left(\beta_2 (x) z_k^2 - \beta_1 (x) y_k^2 \right) dx \to 0$$

as $k \to \infty$, which is impossible.

Next, from (6.30), (6.31) and (6.22),

$$\epsilon \|u\|_{0,2}^2 \leq |\langle f_0 (x, u, \nabla u), z - y \rangle_2| \leq C_1 \|u\|_{0,2}^\theta + C_2 \|u\|_{0,2}$$

for some constants C_1 and C_2; here $\theta = \max \{ \gamma + 1, \eta + 1 \}$. This implies that there exists $r > 0$ independent of λ, with $\|u\|_{0,2} \leq r$. $\square$

Remark 6.5 *In the situation that f does not depend on ∇u, the assumption that Ω is C^2 is not necessary and we get a solution in $H_0^1 (\Omega)$.*

7. Theorems of Leray-Schauder Type for Coincidences

In the previous chapters we were concerned exclusively with the existence of fixed points for an operator T, that is, with the solvability of the equation $x = T(x)$. Here we are concerned with the solvability of more general operator equations of coincidence type, namely $L(x) = T(x)$, where L is a linear operator. We shall extend and complement previous results given in [58], [76] and [143].

7.1 Continuation Principles for Coincidences

Throughout this section, X is a Banach space, Y a normed space and $L : D(L) \subset X \to Y$ a linear *Fredholm map of index zero*, i.e.

$$\text{Im } L \text{ is closed} \quad \text{and} \quad \dim \ker L = \text{codim Im } L < \infty.$$

Let $X = X_1 \oplus X_2$, $Y = Y_1 \oplus Y_2$, where $X_1 = \ker L$ and $Y_2 = \text{Im } L$. Let $P : X \to X_1$, $Q : Y \to Y_1$ be continuous linear projectors and $J : X_1 \to Y_1$ a linear isomorphism. Then $L + JP$ is a bijective linear map.

Let $Z \subset X$ and $T : Z \to Y$. We say that T is *L-compact* (*L-completely continuous*, of *Krasnoselskii L-type, L-set-contraction, L-*

condensing, L-continuous), if

$$(L + JP)^{-1} T : Z \to X$$

is compact (completely continuous, of Krasnoselskii type, set-contraction, condensing, continuous, respectively). Also, we say that T is of *Mönch L-type* if: $(L + JP)^{-1} T$ is continuous; $(L + JP)^{-1} T (Z)$ is bounded when $Y_1 \neq \{0\}$; and there exists $x_0 \in Z$ such that

$$S \subset \overline{\text{conv}} \left\{ \{x_0\} \cup (L + JP)^{-1} T (S) \right\} + K_1 \implies \bar{S} \text{ compact} \qquad (7.1)$$

whenever $S \subset Z$ is countable and $K_1 \subset \ker L$ is compact.

Remark 7.1 *1) The above definitions do not depend on the choice of P, Q and J. This easily follows from the formula*

$$(L + \Phi_1)^{-1} T = (L + \Phi_2)^{-1} T$$

$$+ (L + \Phi_1)^{-1} (\Phi_2 - \Phi_1) (L + \Phi_2)^{-1} T$$

and the complete continuity of

$$(L + \Phi_1)^{-1} (\Phi_2 - \Phi_1) : X \to \ker L \subset X$$

(see [[76], p.39], [[49], p.13]), where Φ_1, $\Phi_2 : X \to Y_1$ are continuous linear maps with $L + \Phi_1$ and $L + \Phi_2$ bijective.

2) L-compact $\implies$ L-completely continuous $\implies$ Krasnoselskii L-type $\implies$ L-set-contraction $\implies$ L-condensing $\implies$ L-continuous; also L-compact $\implies$ Mönch L-type $\implies$ L-continuous and if Z is bounded, then L-condensing $\implies$ Mönch L-type.

3) The map $JP : X \to Y$ is L-completely continuous.

4) If $T : Z \to Y$ is L-completely continuous (of Krasnoselskii L-type, L-set-contraction, L-condensing), then $T + JP$ also has that property.

5) If $D(L)$ is closed (then we may suppose without loss of generality that $D(L) = X$) and L is continuous, then a map is L-completely continuous if and only if it is completely continuous.

Let us consider a closed set $K_0 \subset X$, a nonempty bounded set $U \subset K_0$ open in K_0 and a nonempty convex set $K \subset Y$. Denote by ∂U the boundary of U with respect to K_0.

We can now state the analogue of Theorem 5.3 for coincidences.

Theorem 7.1 *Assume K_0 is convex and*

$$(L + JP)^{-1} (K + JP(\overline{U})) \subset K_0.$$

Let $T : \overline{U} \to K$ be a map of Mönch L-type with (7.1) holding for some $x_0 \in D(L) \cap U$. In addition suppose that

$$Lx \neq (1 - \lambda)(Lx_0 - JP(x - x_0)) + \lambda T(x) \qquad (7.2)$$

for all $x \in D(L) \cap \partial U$ and $\lambda \in [0, 1]$. Then there exists $x \in U$ with $Lx = T(x)$.

Proof. The equation $Lx = T(x)$ is equivalent to

$$x = (L + JP)^{-1} (T + JP)(x).$$

Let

$$\tilde{T} : \overline{U} \to K_0, \quad \tilde{T} = (L + JP)^{-1} (T + JP).$$

We can easily see that $\tilde{T}$ is a Mönch type map which, more exactly, satisfies:

$$S \text{ countable}, \ S \subset \overline{\text{conv}}\{\{x_0\} \cup \tilde{T}(S)\} \implies \overline{S} \text{ compact}.$$

In addition, from (7.2),

$$(1 - \lambda) x_0 + \lambda \tilde{T}(x) \neq x \quad \text{for all } x \in \partial U \text{ and } \lambda \in [0, 1].$$

Thus we may apply Theorem 5.3. $\square$

Remark 7.2 *For $X = Y$, $K_0 = K$ and $L = $ identity of X, Theorem 7.1 reduces to Theorem 5.3.*

Let $\mathbf{M}^L$ be any of the following classes $\mathbf{M}_S^L$, $\mathbf{M}_D^L$, $\mathbf{M}_C^L$ of all L-condensing, L-set-contractions, L-compact maps from $\overline{U}$ into K, respectively, which do not have coincidence points with L on ∂U. A map $T \in \mathbf{M}^L$ is said to be *essential* in $\mathbf{M}^L$ if every $T' \in \mathbf{M}^L$ with $T'(x) = T(x)$ on ∂U has a coincidence point with L.

The next result is the topological transversality theorem for coincidences.

Theorem 7.2 *Let $H : \overline{U} \times [0, 1] \to K$ be L-condensing (L-set-contraction, L-compact, respectively). Assume*

(a) $H(x, \lambda) \neq Lx$ *for all* $x \in \partial U$ *and* $\lambda \in [0, 1]$;

(b) H_0 *is essential in* $\mathbf{M}_S^L$ ($\mathbf{M}_D^L$, $\mathbf{M}_C^L$, *respectively*).

     Then, for each $\lambda \in [0, 1]$, *there exists a coincidence point of* H_λ *with* L. *Moreover,* H_λ *is essential in* $\mathbf{M}_S^L$ ($\mathbf{M}_D^L$, $\mathbf{M}_C^L$, *respectively*) *for every* $\lambda \in [0, 1]$.

Proof.   We use the same reasoning as in the proof of Theorem 5.4. $\square$
     The next proposition gives an example of an essential map.

Proposition 7.3 *Suppose that* K_0 *is convex and*

$$(L + JP)^{-1}(K + JP(\overline{U})) \subset K_0. \tag{7.3}$$

Let $F_0 : \overline{U} \to Y_1$ *be* L-*condensing* (L-*set-contraction,* L-*compact, respectively*) *and let* $x_0 \in D(L) \cap U$. *Assume*

$$Lx_0 + F_0(\overline{U}) \subset K \tag{7.4}$$

and that the following conditions are satisfied:

$$F_0(x) \neq 0 \quad \text{for all } x \in (x_0 + X_1) \cap \partial U \tag{7.5}$$

$$\langle F_0(x), J(x - x_0) \rangle \leq 0 \quad \text{for all } x \in (x_0 + X_1) \cap \partial U. \tag{7.6}$$

where $\langle ., . \rangle$ *denotes the euclidean inner product on* Y_1. *Then the map* $Lx_0 + F_0$ *is essential in* $\mathbf{M}_S^L$ ($\mathbf{M}_D^L$, $\mathbf{M}_C^L$, *respectively*).

Proof.   Let $G \in \mathbf{M}^L$ with $G(x) = Lx_0 + F_0(x)$ on ∂U. We have to prove the solvability in U of the equation $Lx = G(x)$, equivalently

$$x = (L + JP)^{-1}(G + JP)(x).$$

For this, we define the homotopy $H : \overline{U} \times [0, 1] \to K_0$,

$$H(x, \lambda) = (1 - \lambda)x_0 + \lambda(L + JP)^{-1}(G + JP)(x).$$

Since K_0 is convex, H is well defined and, from Remark 7.1 4), H is condensing (set-contraction, compact, respectively) as a map from $\overline{U} \times [0, 1]$ into K_0. In addition,

$$x \neq H(x, \lambda) \quad \text{for all } x \in \partial U \text{ and } \lambda \in [0, 1].$$

To prove this, suppose the contrary, that is suppose $x = H(x, \lambda)$ for some $x \in \partial U$ and $\lambda \in [0, 1]$. If $\lambda = 0$, we have $x = x_0$, a contradiction since $x_0 \in U$. If $\lambda = 1$, then $Lx = G(x)$, equivalently $Lx = Lx_0 + F_0(x)$. Then $L(x - x_0) = 0$ and $F_0(x) = 0$, that is $x \in (x_0 + X_1) \cap \partial U$ and $F_0(x) = 0$, which contradicts (7.5). If $\lambda \in (0, 1)$, then $x - x_0 = \lambda [H(x, 1) - x_0]$. Hence

$$\begin{aligned} (L + JP)(x - x_0) &= \lambda [(G + JP)(x) - (L + JP)x_0] \\ &= \lambda F_0(x) + \lambda JP(x - x_0). \end{aligned}$$

Thus, $L(x - x_0) = 0$, that is $x \in (x_0 + X_1) \cap \partial U$, and

$$\lambda F_0(x) = (1 - \lambda) J(x - x_0).$$

This yields

$$\lambda \langle F_0(x), J(x - x_0) \rangle = (1 - \lambda) |J(x - x_0)|^2 > 0,$$

which contradicts (7.6). Thus our claim is proved. Finally, since H_0 is essential in $\mathbf{M}$, the classical topological transversality theorem (Theorem 5.4) guarantees that H_1 is also essential in $\mathbf{M}$. Thus, there exists $x \in U$ with $x = H_1(x)$, that is $Lx = G(x)$. Hence the map $Lx_0 + F_0$ is essential in $\mathbf{M}^L$. $\square$

Remark 7.3 *1) If*

$$Lx_0 - JP(\overline{U} - x_0) \subset K \tag{7.7}$$

then the map $F_0(x) = -JP(x - x_0)$ satisfies (7.4)-(7.6). Therefore, if K_0 is convex and (7.3), (7.7) hold, then the map $Lx_0 - JP(x - x_0)$ is essential in $\mathbf{M}^L$.

2) If $x_0 \in U \subset K_0 = K \subset X = Y$ and L is the identity of X, conditions (7.3) and (7.7) are satisfied and $Lx_0 - JP(x - x_0)$ is just the constant map x_0.

3) Suppose that $F : \overline{U} \to Y$ is L-compact. Then the map $QF : \overline{U} \to Y_1$ is also L-compact. This follows from the equality

$$(L + JP)^{-1} QF = P(L + JP)^{-1} F.$$

Thus, in Proposition 7.3, we can put $F_0 = QF$, where $F : \overline{U} \to Y$ is any L-compact map.

We now return to the solvability of the operator equation $Lx = T(x)$. If T is L-compact, the following result holds:

Theorem 7.4 *Assume K_0 is convex and condition (7.3) is satisfied. Let $x_0 \in D(L) \cap U$ and $T : \bar{U} \to K$ be L-compact. In addition assume:*

$$Lx_0 + QT(\bar{U}) \subset K \tag{7.8}$$

$$QT(x) \neq 0 \quad \text{for all } x \in (x_0 + X_1) \cap \partial U$$

$$\langle QT(x), J(x - x_0) \rangle \leq 0 \quad \text{for all } x \in (x_0 + X_1) \cap \partial U$$

$$Lx \neq (1 - \lambda) Lx_0 + \lambda T(x) \text{ for } x \in D(L) \cap \partial U, \ \lambda \in (0, 1]. \tag{7.9}$$

Then there exists $x \in D(L) \cap U$ that solves $Lx = T(x)$.

Proof. From Proposition 7.3 and Remark 7.3 3), the map $Lx_0 + QT$ is essential in $\mathbf{M}_C^L$. Suppose the conclusion is false, that is $Lx \neq T(x)$ for all $x \in D(L) \cap U$. Consider the homotopy $H : \bar{U} \times [0, 1] \to K$,

$$H(x, \lambda) = (1 - \lambda)[Lx_0 + QT(x)] + \lambda T(x)$$

which is well defined by (7.8), $T(\bar{U}) \subset K$ and K convex. If $Lx \neq H(x, \lambda)$ for all $x \in D(L) \cap \partial U$ and $\lambda \in [0, 1]$, then since $H_0 = Lx_0 + QT$ is essential in $\mathbf{M}_C^L$, we have from Theorem 7.2 that $H_1 = T$ is essential in $\mathbf{M}_C^L$. This contradicts the assumption $Lx \neq T(x)$. Thus, there exists $x \in D(L) \cap \partial U$ and $\lambda \in [0, 1]$ with $Lx = H(x, \lambda)$. Clearly $\lambda > 0$. On the other hand, the equation $Lx = H(x, \lambda)$ is equivalent to the following system

$$Lx = (1 - \lambda) Lx_0 + \lambda [T(x) - QT(x)], \quad QT(x) = 0.$$

Hence
$$Lx = (1 - \lambda) Lx_0 + \lambda T(x)$$

for some $x \in D(L) \cap \partial U$ and $\lambda \in (0, 1]$, which contradicts (7.9). □

Remark 7.4 *It is clear that, under the hypotheses of Theorem 7.4, for each L-compact map $G : \bar{U} \to K$ satisfying $Lx_0 + QG(\bar{U}) \subset K$ and $G(x) = T(x)$ on ∂U, there exists an $x \in D(L) \cap U$ with $Lx = G(x)$. Thus, if in Theorem 7.4 $K = Y$, then T is essential in $\mathbf{M}_C^L$.*

7.2 Application to Periodic Solutions of Differential Systems

The differential operators

$$L : C_B^m \to C, \quad Lu = u^{(m)}$$

and

$$L : C_B^{m-1} \to C_0, \quad Lu = u^{(m-1)} - u^{(m-1)}(0)$$

are not invertible if $\mathcal{B}$ represents the periodic conditions

$$\mathcal{B} : u^{(k)}(0) = u^{(k)}(1), \quad k = 0, 1, ..., m - 1.$$

Therefore, a natural application of the general coincidences results stated above is periodic problems.

For example, let us consider the periodic boundary value problem for a system of n first order differential equations

$$\begin{cases} u' = f(t, u), & t \in I \\ u(0) = u(1) \end{cases} \tag{7.10}$$

Let $L : C_B \to C_0$ be given by

$$(Lu)(t) = u(t) - u(0).$$

It is elementary to check that

$$\ker L = \{ u \in C_B; \ u(t) \equiv a, \ a \in \mathbf{R}^n \}$$

and

$$\operatorname{Im} L = \{ v \in C_0; \ v(1) = 0 \}.$$

Since each $v \in C_0$ can be expressed as

$$v(t) = tv(1) + [v(t) - tv(1)],$$

we have the direct sum decomposition $C_0 = (t\mathbf{R}^n) \oplus \operatorname{Im} L$. Thus, L is a Fredholm map of index zero and we may set

$$Pu = u(1), \quad (Qv)(t) = tv(1) \quad \text{and} \quad (Ja)(t) = ta.$$

Assume $f : I \times \overline{B}_R(0; \mathbf{R}^n) \to \mathbf{R}^n$ is a continuous function. Then we may define $T : D \to C_0$, by

$$T(u)(t) = \int_0^t f(\tau, u(\tau)) \, d\tau$$

where $D = \{u \in C_B ; \ |u(t)| \leq R \text{ for all } t \in I\}$. It is well known that T is a compact map. In this case, the map $(L + JP)^{-1}$ is continuous and consequently, T is L-compact.

Lets introduce the family of periodic problems

$$\begin{cases} u' = \lambda f(t, u), & t \in I \\ u(0) = u(1) \end{cases} \tag{7.11}$$

where $\lambda \in [0, 1]$.

We have the following existence principle.

Theorem 7.5 *Let $f : I \times \overline{B}_R(0; \mathbf{R}^n) \to \mathbf{R}^n$ be a continuous function. Assume that the following conditions are satisfied:*

(A) *for all $a \in \mathbf{R}^n$ with $|a| = R$, one has*

$$\int_0^1 f(s, a) \, ds \neq 0 \quad \text{and} \quad \langle a, \int_0^1 f(s, a) \, ds \rangle \leq 0;$$

(B) *$|u(t)| < R$ for all $t \in I$ and any solution $u \in C^1(I; \mathbf{R}^n)$ to (7.11), where $\lambda \in [0, 1]$.*

Then there exists $u \in C^1(I; \mathbf{R}^n)$ that solves (7.10).

Proof. Apply Theorem 7.4. $\square$

We now deduce an existence result (compare with Theorem 10.1 in [58]).

Theorem 7.6 *Let $f : I \times \mathbf{R}^n \to \mathbf{R}^n$ be a continuous function. Assume that the following conditions are satisfied:*

(a) *there exists a sequence (R_k) of positive numbers tending to infinity such that*

$$\int_0^1 f(s, a) \, ds \neq 0, \quad \langle a, \int_0^1 f(s, a) \, ds \rangle \leq 0$$

for all $a \in \mathbf{R}^n$ with $|a| = R_k$ and every k;

(b) *there is an $r > 0$ such that $|a| \le r$ for all $a \in \mathbf{R}^n$ satisfying*
   $\langle a, f(t, a) \rangle = 0$ *for some $t \in I$;*

(c) *there exists $\beta \in C(I; \mathbf{R}_+)$, $\psi \in C(\mathbf{R}_+; \mathbf{R}_+)$ and $r_1 > r$ such that*

$$|f(t, u)| \le \beta(t) \psi(|u|) \quad \text{for all } t \in I, \ u \in \mathbf{R}^n$$

$$\int_{r_1}^{\infty} \frac{d\tau}{\psi(\tau)} > \int_r^{r_1} \frac{d\tau}{\psi(\tau)} = \int_0^1 \beta(\tau) \, d\tau.$$

Then there exists $u \in C^1(I; \mathbf{R}^n)$ that solves (7.10).

Proof. Let u be any solution of (7.11) for some $\lambda \in (0, 1]$. We deduce that

$$0 = |u(1)|^2 - |u(0)|^2 = 2\lambda \int_0^1 \langle u(s), f(s, u(s)) \rangle \, ds.$$

Thus

$$\langle u(t), f(t, u(t)) \rangle = 0 \text{ for some } t_u \in I.$$

Now (b) guarantees $|u(t_u)| \le r$. If $|u(0)| = |u(1)| > r$, then there is a largest $c < 1$ such that $|u(c)| = r$ and $|u(t)| > r$ on $(c, 1]$. Then, on $(c, 1]$, we have

$$|u(s)|' \le |u'(s)| = \lambda |f(s, u(s))| \le \beta(s) \psi(|u(s)|).$$

It follows that

$$\int_r^{|u(1)|} \frac{d\tau}{\psi(\tau)} = \int_c^1 \frac{|u(s)|'}{\psi(|u(s)|)} ds \le \int_0^1 \beta(\tau) \, d\tau.$$

This with (c), implies that $|u(1)| \le r_1$. Thus,

$$|u(0)| = |u(1)| \le r_1.$$

Now suppose that $|u(t)| > r_1$ for some t. Then there is a largest $c' < t$ such that $|u(c')| = r_1$ and $|u(s)| > r_1$ on $(c', t]$. By a similar reasoning we get a constant $r_2 \ge r_1$ independent of u, with $|u(t)| \le r_2$. Finally, we choose $R = R_k > r_2$ and we apply Theorem 7.5. $\square$

We might think of extending Theorems 7.5 and 7.6 to equations in Banach spaces. Unfortunately, the theory above does not work since $\ker L$ and Y_1 are not necessarily finite dimensional.

Remark 7.5 *In the case when X_1 is infinite dimensional but Y_1 is finite dimensional (L is said to be semi-Fredholm to the right), the theory can easily be extended as was shown by Krawcewicz [76]: Assume $L : D(L) \subset X \to Y$ is a linear map satisfying the following hypotheses:*

(1) L is a closed map;

(2) ker L and Im L are closed and there are two closed subspaces $X_2 \subset X$ and $Y_1 \subset Y$ such that $X = X_1 \oplus X_2$ and $Y = Y_1 \oplus Y_2$, where $X_1 = ker L$ and $Y_2 = Im L$;

(3) Y_1 is finite dimensional.

Then, since L is a closed map and Im L is a closed space, the map $L_{X_2} : D(L) \cap X_2 \to Y_2$ is invertible and its inverse $R : Y_2 \to X_2$ is a linear continuous map. Then the equation $Lx = T(x)$, $x \in X_2$ is equivalent to the system

$$x = R(I - Q)T(x), \quad 0 = QT(x),$$

where Q is the linear projector into Y_1 associated with the decomposition $Y = Y_1 \oplus Y_2$. If L is surjective, that is $Q = 0$, then the equation is equivalent to $x = RT(x)$. This equation can be discussed by means of the methods already seen assuming, for example, that RT is condensing. The case when L is not surjective can always be reduced to the previous one by taking a linear continuous map $S : X \to Y_1$ (hence completely continuous since its range is finite dimensional) such that $L + S$ is surjective.

Related topics and applications can be found in [40], [49], [51], [65], [75], [88], [134] and [158]. Approximation-solvability of coincidence equations of the form $Lx = T(x)$ was extensively studied by Petryshyn (see [122] and [123]) using the so called A-proper technique when dim ker $L < \infty$. The case when ker L is infinite dimentional has also been studied (see [92] and [118]).

8. Theorems of Selective Continuation

In Chapters 2-7, we were concerned exclusively with homotopies $H(x, \lambda)$ defined on a cylinder $\overline{\mathcal{U}} = \overline{U} \times [0, 1]$. Thus, all operators $H_\lambda = H(,,\lambda)$ $(\lambda \in [0, 1])$ had the same domain $\overline{U}$. Here we are concerned more generally with homotopies H for which the operators H_λ may have different domains $\overline{U}_\lambda$. This situation arises when we look for solutions having a particular property. The idea is to try to follow a branch of solutions to $H_\lambda(x) = x$ with the desired property and thus to work on some neighborhood $\mathcal{U}$ of that branch which avoids all the other solutions. This kind of continuation with solutions having a particular property, will be called *selective continuation*.

8.1   Selective Continuation Principles

The main idea of selective continuation is to reduce the study of the family $\{H_\lambda\}$ to that of a certain family of maps from the same domain $\overline{\mathcal{U}}$ into $Y \times \mathbf{R}$. Thus we pass from maps acting between the spaces X and Y, to maps acting between the product spaces $X \times \mathbf{R}$ and $Y \times \mathbf{R}$. Such an idea has been used by Furi and Pera [48] and Fitzpatrick, Massabò and Pejsachowicz [41] and was successful in proving the solvability of boundary value problems when a priori bounds of solutions can be obtained only for the solutions having the particular property.

Throughout this section we use the notation of Chapter 7. In addition we let $\mathcal{X} = X \times \mathbf{R}$, $\mathcal{Y} = Y \times \mathbf{R}$, $\mathcal{K}_0 = K_0 \times [0, 1]$, $\mathcal{K} = K \times [0, 1]$

and define $\mathcal{L} : D(\mathcal{L}) \subset \mathcal{X} \to \mathcal{Y}$, where $D(\mathcal{L}) = D(L) \times \mathbf{R}$, by

$$\mathcal{L}(x, \lambda) = (Lx, \lambda).$$

It is easy to check that $\mathcal{L}$ is a linear Fredholm map of index zero and

$$\ker \mathcal{L} = X_1 \times \{0\}, \quad \operatorname{Im} \mathcal{L} = Y_2 \times \mathbf{R}.$$

Furthermore, let us consider

$$\mathcal{P} : \mathcal{X} \to X_1 \times \{0\}, \quad \mathcal{P}(x, \lambda) = (Px, 0),$$

$$\mathcal{Q} : \mathcal{Y} \to Y_1 \times \{0\}, \quad \mathcal{Q}(x, \lambda) = (Qx, 0)$$

and

$$\mathcal{J} : X_1 \times \{0\} \to Y_1 \times \{0\}, \quad \mathcal{J}(x, 0) = (Jx, 0).$$

Notice that

$$(\mathcal{L} + \mathcal{JP})^{-1}(y, \lambda) = ((L + JP)^{-1}y, \lambda) \quad \text{for every } (y, \lambda) \in \mathcal{Y}.$$

For any $\mathcal{V} \subset X \times [0,1]$, we denote by $\mathcal{V}_\lambda = \{x \in X; \ (x, \lambda) \in \mathcal{V}\}$ the section of $\mathcal{V}$ at λ.

Let $\mathcal{U} \subset \mathcal{K}_0$ be nonempty open and bounded and $\partial \mathcal{U}$ the boundary of $\mathcal{U}$ in $\mathcal{K}_0$. Denote by $\mathcal{M}_S^{\mathcal{L}}$, $\mathcal{M}_D^{\mathcal{L}}$, $\mathcal{M}_C^{\mathcal{L}}$ the classes of maps $\mathbf{M}_S^{\mathcal{L}}(\overline{\mathcal{U}}; \mathcal{K})$, $\mathbf{M}_D^{\mathcal{L}}(\overline{\mathcal{U}}; \mathcal{K})$ and $\mathbf{M}_C^{\mathcal{L}}(\overline{\mathcal{U}}; \mathcal{K})$, respectively.

We can extend Theorem 7.2 as follows.

Theorem 8.1 *Let* $H : \overline{\mathcal{U}} \to K$ *be* L-*condensing* (L-*set-contraction,* L-*compact). Assume*

(i) $H(x, \lambda) \neq Lx$ *for all* $(x, \lambda) \in \partial \mathcal{U}$;

(ii) *the map* $\mathcal{H}_0(x, \lambda) := (H(x, \lambda), 0)$ *is essential in* $\mathcal{M}_S^{\mathcal{L}}$ ($\mathcal{M}_D^{\mathcal{L}}$, $\mathcal{M}_C^{\mathcal{L}}$, *respectively).*

Then, for each $\lambda \in [0,1]$, *there exists in* $\mathcal{U}_\lambda$ *a coincidence point of* H_λ *with* L. *Moreover, the map*

$$\mathcal{H}_\mu : \overline{\mathcal{U}} \to \mathcal{K}, \quad \mathcal{H}_\mu(x, \lambda) = (H(x, \lambda), \mu)$$

is essential in $\mathcal{M}_S^{\mathcal{L}}$ ($\mathcal{M}_D^{\mathcal{L}}$, $\mathcal{M}_C^{\mathcal{L}}$, *respectively) for every* $\mu \in [0,1]$.

Proof. We apply Theorem 7.2 to $\mathcal{X}$, $\mathcal{Y}$, $\mathcal{K}_0$, $\mathcal{K}$, $\mathcal{U}$, $\mathcal{L}$, $\mathcal{M}^{\mathcal{L}}$ and $\mathcal{H}$ instead of X, Y, K_0, K, U, L, $\mathbf{M}^L$ and H, where

$$\mathcal{H} : \overline{\mathcal{U}} \times [0,1] \to \mathcal{K},$$
$$\mathcal{H}(x,\lambda,\mu) = (H(x,\lambda),\mu) \quad \text{for } (x,\lambda) \in \overline{\mathcal{U}}, \ \mu \in [0,1].$$

From assumption (i), we can easily check that

$$\mathcal{H}(x,\lambda,\mu) \neq \mathcal{L}(x,\lambda) \quad \text{for all } (x,\lambda) \in \partial\mathcal{U}, \ \mu \in [0,1].$$

Thus $\mathcal{H}$ satisfies both conditions (a) and (b) of Theorem 7.2. Notice that if $(x,\lambda) \in \mathcal{U}$ is a coincidence point of $\mathcal{H}_\mu$ and $\mathcal{L}$, then $H(x,\lambda) = Lx$ and $\mu = \lambda$, and so, $x \in \mathcal{U}_\mu$ and $H(x,\mu) = Lx$. $\square$

Remark 8.1 *In the case when $\mathcal{U}$ is of the form $\mathcal{U} = U \times [0,1]$, where U is a bounded open subset of K_0, (ii) implies condition (b) of Theorem 7.2. Indeed, if $T \in \mathbf{M}^L = \mathbf{M}^L(\overline{U}; K)$ and T and H_0 coincide on ∂U, then the map $\mathcal{T}(x,\lambda) := (T(x),0)$ is in $\mathcal{M}^{\mathcal{L}} = \mathbf{M}^L(\overline{\mathcal{U}}; \mathcal{K})$ and $\mathcal{T}(x,\lambda) = (H(x,0),0)$ for all $(x,\lambda) \in \partial\mathcal{U}$. On the other hand, the maps $(H(x,0),0)$ and $(H(x,\lambda),0)$ are homotopic via the homotopy*

$$(x,\lambda,\mu) \in \overline{\mathcal{U}} \times [0,1] \longmapsto (H(x,\mu\lambda),0).$$

Thus, from (ii), the map $(H(x,0),0)$ is essential in $\mathcal{M}^{\mathcal{L}}$, and so $\mathcal{T}$ is also essential in $\mathcal{M}^{\mathcal{L}}$. Consequently, $\mathcal{T}$ and $\mathcal{L}$ have a coincidence point in $\mathcal{U}$, that is T and L have a coincidence point in U. Therefore, H_0 is essential in $\mathbf{M}^L$ as claimed.

The next result is concerned with a sufficient condition for (ii) to hold, namely that H_0 be homotopic on $\mathcal{U}_0$ to a map of the form $Lx_0 + F_0(x)$ like that in Proposition 7.3.

Theorem 8.2 *Suppose that K_0 is convex and*

$$(L+JP)^{-1}(K + JP(K_0)) \subset K_0. \tag{8.1}$$

Let $F_0 : K_0 \to Y_1$ be L-condensing (L-set-contraction, L-completely continuous, respectively) and $x_0 \in D(L) \cap \mathcal{U}_0$. Assume that the following conditions are satisfied:

$$Lx_0 + F_0(K_0) \subset K \tag{8.2}$$

$$F_0(x) \neq 0 \quad \text{for all } x \in (x_0 + X_1) \text{ with } (x, 0) \in \partial \mathcal{U} \qquad (8.3)$$

$$\langle F_0(x), J(x - x_0) \rangle \leq 0 \text{ for all } x \in (x_0 + X_1) \text{ with } (x, 0) \in \partial \mathcal{U}. \quad (8.4)$$

If $H : \overline{\mathcal{U}} \to K$ *is* L-condensing (L-set-contraction, L-compact, respectively), satisfies (i) and

$$Lx \neq (1 - \mu)(Lx_0 + F_0(x)) + \mu H(x, 0) \qquad (8.5)$$

for all $(x, 0) \in \partial \mathcal{U}$, $\mu \in (0, 1)$, *then there exists* $x \in D(L) \cap \mathcal{U}_1$ *with* $Lx = H(x, 1)$.

Proof. We show that (ii) is satisfied and then we apply Theorem 8.1. For this, we consider the homotopy

$$\widetilde{\mathcal{H}} : \overline{\mathcal{U}} \times [0, 1] \to \mathcal{K},$$

$$\widetilde{\mathcal{H}}(x, \lambda, \mu) = ((1 - \mu)(Lx_0 + F_0(x)) + \mu H(x, \lambda), 0).$$

It is immediate that $\widetilde{\mathcal{H}}$ is $\mathcal{L}$-condensing ($\mathcal{L}$-set-contraction, $\mathcal{L}$-compact, respectively). Also,

$$\mathcal{L}(x, \lambda) \neq \widetilde{\mathcal{H}}(x, \lambda, \mu) \quad \text{for all } (x, \lambda) \in \partial \mathcal{U} \text{ and } \mu \in [0, 1]$$

(use (8.5) when $\mu \in (0, 1)$, (8.3) when $\mu = 0$ and (i) when $\mu = 1$).
  For $\mu = 0$,

$$\widetilde{\mathcal{H}}(x, \lambda) = (Lx_0 + F_0(x), 0) = \mathcal{L}(x_0, 0) + \mathcal{F}_0(x, \lambda),$$

where

$$\mathcal{F}_0 : \overline{\mathcal{U}} \to Y_1 \times \{0\}, \quad \mathcal{F}_0(x, \lambda) = (F_0(x), 0).$$

Now we can easily check that all the hypotheses of Proposition 7.3 are satisfied for

$$\mathcal{X}, \ \mathcal{Y}, \ \mathcal{K}_0, \ \mathcal{K}, \ \mathcal{U}, \ \mathcal{L}, \ \mathcal{J}, \ \mathcal{P}, \ \mathcal{F}_0, \ \mathcal{M}^{\mathcal{L}} \text{ and } (x_0, 0)$$

instead of

$$X, \ Y, \ K_0, \ K, \ U, \ L, \ J, \ P, \ F_0, \ \mathbf{M}^L \text{ and } x_0.$$

It follows that $\widetilde{\mathcal{H}}_0$ is essential in $\mathcal{M}^{\mathcal{L}}$ and so, by Theorem 7.2, $\widetilde{\mathcal{H}}_1 = \mathcal{H}_0$ is essential in $\mathcal{M}^{\mathcal{L}}$ too. Thus, (ii) holds and Theorem 8.1 applies. $\square$

Remark 8.2 *1) For $K_0 = X$ and $K = Y$, conditions (8.1) and (8.2) trivially hold.*

2) The map $F_0(x) = -JP(x - x_0)$ satisfies (8.3) and (8.4).

Notice that for $X = Y$, $K_0 = K$ and $L =$ identity of X, Theorem 8.2 becomes Corollary 1 in [138] (in this case $F_0 = 0$).

The next result is the analogue of Theorem 7.4 in this situation.

Theorem 8.3 *Assume K_0 is convex and that (8.1) holds. Let $x_0 \in D(L) \cap \mathcal{U}_0$ and $T : K_0 \to K$ be L-completely continuous. In addition suppose that the following conditions are satisfied:*

$$QT(x) \neq 0 \quad \text{for } x \in x_0 + X_1 \text{ with } (x, 0) \in \partial \mathcal{U}$$

$$\langle QT(x), J(x - x_0) \rangle \leq 0 \quad \text{for } x \in x_0 + X_1 \text{ with } (x, 0) \in \partial \mathcal{U}$$

$$Lx \neq (1 - \lambda) Lx_0 + \lambda T(x) \quad \text{for } (x, \lambda) \in \partial \mathcal{U} \text{ with } \lambda \in (0, 1].$$

Then there exists $x \in D(L) \cap \mathcal{U}_1$ with $Lx = T(x)$.

Proof. Check that all assumptions of Theorem 7.4 are satisfied for

$$\mathcal{X}, \ \mathcal{Y}, \ \mathcal{K}_0, \ \mathcal{K}, \ \mathcal{U}, \ \mathcal{L}, \ \mathcal{J}, \ \mathcal{P}, \ \mathcal{Q}, \ \mathcal{T} \text{ and } (x_0, 0)$$

instead of

$$X, \ Y, \ K_0, \ K, \ U, \ L, \ J, \ P, \ Q, \ T \text{ and } x_0,$$

where

$$\mathcal{T} : \overline{\mathcal{U}} \to \mathcal{K}, \quad \mathcal{T}(x, \lambda) = (T(x), 1).$$

Then apply Theorem 7.4. $\square$

For similar results in terms of topological degree we refer to [24], [89].

8.2 Continua of Solutions

The continuation theorems proved in Chapter 2 for contraction mappings show that the solution set $\{(x, \lambda) ; \ H(x, \lambda) = x\}$ is a compact and connected (continuum) set joining the point $(x(0), 0)$ to $(x(1), 1)$, where $x(0)$ is the fixed point of H_0 and $x(1)$ is the fixed point of H_1. The goal of this section is to obtain a similar result for the continuation theorems proved in this chapter.

Let (K, d) be a metric space. A set $M \subset K$ is said to be *connected*if it can not be represented as

$$M = M_0 \cup M_1 \text{ with } M_0 \cap \overline{M}_1 = \overline{M}_0 \cap M_1 = \emptyset.$$

A set $M \subset K$ is said to be *well-chained* provided that for every $\varepsilon > 0$, any two points $a, b \in M$ can be joined by an ε-chain of points all lying in M. An ε-chain joining a and b is a finite sequence of points $a = x_1, x_2, ..., x_n = b$ such that

$$d(x_k, x_{k+1}) < \varepsilon \text{ for } k = 1, 2, ..., n - 1.$$

It is easy to show that a compact set is connected if and only if it is well-chained. A compact connected set will be called a *continuum*.

The next result from general topology is known as Whyburn's lemma.

Lemma 8.4 *Let A and B be disjoint closed subsets of a compact metric space (K, d) such that no connected component of K intersects both A and B. Then there exists a partition $K = K_1 \cup K_2$, where K_1 and K_2 are disjoint compact sets containing A and B, respectively.*

Proof. We first show that there exists an $\varepsilon > 0$ such that no ε-chain in K joins a point of A to a point of B. If this is not so, then for each positive integer n, there exists an $1/n$-chain A_n in K joining a point $a_n \in A$ to a point $b_n \in B$. Since the sequence $(a_n)_{n \geq 1}$ has a convergent subsequence, we may assume that the whole sequence converges. Suppose

$$A_n = \{x_{n,m}; m = 1, 2, ..., m_n\}, \quad a_n = x_{n,1}.$$

Let

$$M = \overline{\bigcup_{n \geq 1} A_n}.$$

It is clear that M is a compact subset of K. Let us prove that M is well-chained. Let $\varepsilon > 0$ be arbitrarily fixed and let $a, b \in M$. Since $(a_n)_{n \geq 1}$ is convergent, there exists $n_\varepsilon \geq 1/\varepsilon$ with

$$d(a_n, a_m) < \varepsilon \text{ for } n \geq n_\varepsilon, m \geq n_\varepsilon.$$

Furthermore, since a is a limit point of $\cup A_n$ and each A_n is finite, there exists $i \geq n_\varepsilon$ and a k_i with $d(a, x_{i,k_i}) < \varepsilon$. Similarly, there exists

$j \geq n_\varepsilon$ and a k_j with $d(b, x_{j,k_j}) < \varepsilon$. It is clear that the following finite sequence

$$a, \; x_{i,k_i}, \; x_{i,k_i-1}, \; ..., \; x_{i,1} = a_i, \; a_j = x_{j,1}, \; x_{j,2}, \; ..., \; x_{j,k_j-1}, \; x_{j,k_j}, \; b$$

is an ε-chain in M joining a to b. Hence M is a connected component of K. But since $a_n \in A \cap A_n$, $b_n \in B \cap A_n$, and A and B are compact, we have $A \cap M \neq \emptyset \neq B \cap M$, contrary to the assumption that no connected component of K intersects both A and B. Thus an ε with the desired property exists.

Now let K_1 be the set of all points of K which can be joined to some point of A by an ε-chain in K and let $K_2 = K \setminus K_1$. It is clear that $A \subset K_1$ and $B \subset K_2$. Now, we claim that K_1 and K_2 are compact sets. To prove this claim it suffices to show that K_1 is both open and closed in K. Indeed, let $x_0 \in K_1$. Then, there is an ε-chain C joining x_0 to some element of A. Consequently, any element $x \in K$ satisfying $d(x_0, x) < \varepsilon$ also belongs to K_1, because the set $C \cup \{x\}$ is an ε-chain joining x to a point of A. Hence K_1 is open. Finally, suppose that

$$(x_k) \subset K_1 \quad \text{and} \quad x_k \to x \in K \text{ as } k \to \infty.$$

Then, there is a k sufficiently large with $d(x_k, x) < \varepsilon$. Since $x_k \in K_1$, there exists an ε-chain C in K joining x_k to some point of A. Clearly, $C \cup \{x\}$ is an ε-chain and so $x \in K_1$. This shows that K_1 is closed and the lemma is proved. $\square$

We now state and prove a more general version of Theorem 8.1.

Theorem 8.5 *Suppose that all the assumptions of Theorem 8.1 are satisfied. Let*

$$\Sigma = \left\{ (x, \lambda) \in \overline{U}; \; Lx = H(x, \lambda) \right\}$$

and, for each $\lambda \in [0, 1]$, let

$$\Sigma_\lambda = \{ x \in X; \; (x, \lambda) \in \Sigma \}.$$

Then Σ contains a continuum intersecting $\Sigma_0 \times \{0\}$ and $\Sigma_1 \times \{1\}$.

Proof. Clearly Σ, $A = \Sigma_0 \times \{0\}$ and $B = \Sigma_1 \times \{1\}$ are compact sets, $A \subset \Sigma$ and $B \subset \Sigma$. If there is no continuum intersecting A and B, then it follows from Whyburn's lemma that Σ can be represented as $\Sigma = \Sigma' \cup \Sigma''$, where Σ' and Σ'' are disjoint compact sets and $A \subset \Sigma'$,

$B \subset \Sigma''$. By Urysohn's lemma, there exists a function $v \in C(\overline{\mathcal{U}}; [0,1])$ such that

$$v(x, \lambda) = 0 \text{ on } \Sigma'' \cup \partial\mathcal{U} \text{ and } v(x, \lambda) = 1 \text{ on } \Sigma'.$$

The map $T(x, \lambda) = (H(x, \lambda), v(x, \lambda))$ belongs to class $\mathcal{M}^{\mathcal{L}}$ and coincides with $\mathcal{H}_0$ on ∂U. Since $\mathcal{H}_0$ is essential, there exists $(x, \lambda) \in \mathcal{U}$ with $T(x, \lambda) = \mathcal{L}(x, \lambda)$. Hence

$$H(x, \lambda) = Lx, \quad v(x, \lambda) = \lambda.$$

If $(x, \lambda) \in \Sigma'$, then $v(x, \lambda) = 1$. Consequently, $\lambda = 1$ and $H(x, 1) = Lx$. Thus, we have $(x, 1) \in B \subset \Sigma''$, a contradiction. Similarly, if $(x, \lambda) \in \Sigma''$, then $v(x, \lambda) = 0$. So $\lambda = 0$ and $H(x, 0) = Lx$, whence $(x, 0) \in A \subset \Sigma'$, a contradiction. $\square$

8.3   Continuation with Respect to a Functional

To make the above results practical we need to construct a set $\mathcal{U}$ with the desired properties. Such a construction was first described by Capietto, Mawhin and Zanolin [23] using a continuous functional on $\mathcal{K}_0$. Then, roughly speaking, $\mathcal{U}$ will be a level set of that functional. The results in [23] are stated in the framework of coincidence degree theory. In [138], an approach without degree was presented for fixed point problems, and in [143], the results were extended to coincidences.

Suppose K_0 is convex, $x_0 \in K_0$, $F_0 : K_0 \to Y_1$ is L-condensing and that conditions (8.1) and (8.2) hold. In addition suppose that

$$F_0(x) \neq 0 \quad \text{for all } x \in (x_0 + X_1) \text{ with } x \neq x_0 \qquad (8.6)$$

$$\langle F_0(x), J(x - x_0) \rangle \leq 0 \quad \text{for all } x \in x_0 + X_1. \qquad (8.7)$$

Let $H : K_0 \times [0,1] \to K$ be L-condensing and let

$$\Sigma = \{(x, \lambda) \in K_0 \times [0,1]; Lx = H(x, \lambda)\},$$

$$\Sigma(x_0) = \{(x, 0); x \in K_0 \text{ and } Lx = (1 - \mu)(Lx_0 + F_0(x))$$

$$+ \mu H(x, 0) \text{ for some } \mu \in [0,1]\}.$$

Also consider a continuous functional $\Phi : K_0 \times [0,1] \to \mathbf{R}$.

Theorem 8.6 *Assume that there are constants c_- and c_+, $c_- < c_+$, such that if we let $\mathcal{V} = \Phi^{-1}((c_-, c_+))$, the following conditions are satisfied:*

(i1) $\Sigma \cap \mathcal{V}$ *is bounded;*

(i2) $\Phi(\Sigma) \cap \{c_-, c_+\} = \emptyset$;

(i3) $\Sigma(x_0)$ *is bounded and included in $\mathcal{V}$.*

Then there exists $x \in D(L) \cap \mathcal{V}_1$ with $Lx = H(x, 1)$.

Proof. Let $\Sigma^* = \Sigma \cap \Phi^{-1}([c_-, c_+])$. From (i2), $\Sigma^* = \Sigma \cap \mathcal{V}$. Also (i1) and the continuity of Φ implies that Σ^* is compact. Hence Σ^* is a compact set included in the open set $\mathcal{V}$. Thus, there exists a bounded open set $\mathcal{U}'$ of $\mathcal{K}_0$ with

$$\Sigma^* \subset \mathcal{U}' \subset \overline{\mathcal{U}'} \subset \mathcal{V}.$$

On the other hand, from (i3), $\Sigma(x_0)$ is another compact set included in $\mathcal{V}$. Thus, there exists a bounded open set $\mathcal{U}''$ of $\mathcal{K}_0$ with

$$\Sigma(x_0) \subset \mathcal{U}'' \subset \overline{\mathcal{U}''} \subset \mathcal{V}.$$

Now the conclusion follows from Theorem 8.2 with $\mathcal{U} = \mathcal{U}' \cup \mathcal{U}''$. $\square$

The functional Φ is said to be *proper on* Σ provided that $\Sigma \cap \Phi^{-1}((a, b))$ is bounded (equivalently, relatively compact) for each bounded real interval (a, b).

Corollary 8.7 *Suppose that the following conditions are satisfied:*

(i1') Φ *is proper on Σ;*

(i2') Φ *is lower bounded on Σ and there is a sequence (c_k) of real numbers with $c_k \to \infty$ as $k \to \infty$ and $c_k \notin \Phi(\Sigma)$ for all k;*

(i3') $\Sigma(x_0)$ *is bounded.*

Then there exists $x \in D(L) \cap \mathcal{K}_0$ with $Lx = H(x, 1)$.

Proof. Now (i3') and the fact that F_0 and H are L-condensing guarantees that the set $\Sigma(x_0)$ is in fact compact. Since Φ is continuous, there are constants a and b with

$$a < \Phi(x, \lambda) < b \text{ for all } (x, \lambda) \in \Sigma(x_0).$$

Furthermore, using (i2'), we can choose c_- and k sufficiently large so that

$$c_- \leq a, \quad c_- < \inf\{\Phi(x, \lambda); (x, \lambda) \in \Sigma\}, \quad c_+ = c_k \geq b.$$

Now we can easily check (i1)-(i3) and we apply Theorem 8.6. □

8.4 Periodic Solutions of Superlinear Singular Boundary Value Problems

In [145], we considered the existence of periodic solutions of superlinear singular equations of the form

$$\frac{1}{p}(pu')' = -g(u) + f(t, u, pu'), \quad t \in (0, 1) \tag{8.8}$$

where

(h1) $p \in C[0, 1] \cap C^1(0, 1)$, $p > 0$ on $(0, 1)$, $g \in C(\mathbf{R})$,

$$g(u)/u \to \infty \text{ as } |u| \to \infty,$$

$f : [0, 1] \times \mathbf{R}^2 \to \mathbf{R}$ is a Carathéodory function and

$$|f(t, u, v)| \leq c(|u| + |v|) + k(t)$$

for all $u, v \in \mathbf{R}$ and a.e. $t \in (0, 1)$, where $c \geq 0$;

(h2) $1/p \in L^1(0, 1)$ and $k \in L^1(0, 1)$.

Without loss of generality, we suppose that

$$p \leq 1 \text{ on } [0, 1] \quad \text{and} \quad ug(u) > 0 \text{ for } u \neq 0.$$

By a periodic solution of (8.8) we mean a function $u \in C[0, 1] \cap C^1(0, 1)$ with $pu' \in C[0, 1]$, differentiable a.e. on $(0, 1)$, which satisfies (8.8) a.e. on $[0, 1]$, and the periodic conditions

$$u(0) = u(1), \quad (pu')(0) = (pu')(1). \tag{8.9}$$

Because we do not require $p(0)p(1) \neq 0$, equation (8.8) may be *singular* at $t = 0$ or $t = 1$.

Singular two point boundary value problems have been studied by a variety of authors, see [102] and the references therein. The novelty here is that the nonlinearity is *superlinear*, i.e. $g(u)/u \to \infty$ as $|u| \to \infty$ and that the set of all solutions may not be bounded. The regular case $p \equiv 1$ has been discussed by Capietto, Mawhin and Zanolin [23].

We shall apply Corollary 8.7. As spaces X and Y we take respectively

$$C_B^1 = \{u \in C[0,1] \cap C^1(0,1); \ u \text{ satisfies } (8.9), \ pu' \in C[0,1]\}$$

with norm $\|u\| = (\|u\|_\infty^2 + \|pu'\|_\infty^2)^{1/2}$ and,

$$C_0 = \{u \in C[0,1]; \ u(0) = 0\}$$

with the usual norm $\|u\|_\infty$. Recall that $\|.\|_q$ stands for the norm of the space $L^q(0,1)$ for $1 \leq q \leq \infty$.

We define $L : C_B^1 \to C_0$ by

$$(Lu)(t) = (pu')(t) - (pu')(0).$$

As in Section 7.2, we can check that

$$\ker L = \left\{u \in C_B^1; \ u(t) = a \text{ on } [0,1], \ a \in \mathbf{R}\right\},$$

$$\operatorname{Im} L = \{v \in C_0; \ v(1) = 0\}$$

and $C_0 = (t\mathbf{R}) \oplus \operatorname{Im} L$. Thus, L is a Fredholm map of index zero and we may set

$$Pu = u(1), \quad (Qv)(t) = tv(1) \quad \text{and} \quad (Ja)(t) = ta.$$

We also consider the map $H : C_B^1 \times [0,1] \to C_0$, given by

$$H(u, \lambda)(t) =$$

$$\int_0^t \left\{ -(1-\lambda)\left[\frac{1}{p}g(u) + \frac{p^2 u'}{1+|pu'|}\right] - \lambda p g(u) + \lambda p f(s, u, pu') \right\} ds.$$

From (h1), (h2) and the Arzelà-Ascoli theorem, H is completely continuous on $C_B^1 \times [0,1]$ and, since in our case $(L + JP)^{-1}$ is continuous,

H is also L-completely continuous. Observe that to solve (8.8)-(8.9) we have to find a $u \in C_B^1$ with $Lu = H(u, 1)$.

Next we define $F_0 : C_B^1 \to t\mathbf{R}$, as

$$F_0(u)(t) = -t \int_0^1 \left(\frac{1}{p} g(u) + \frac{p^2 u'}{1 + |pu'|} \right) ds.$$

Clearly, F_0 is completely continuous and, consequently, L-completely continuous. Also, since $zg(z) > 0$ for $z \neq 0$, F_0 satisfies (8.6) with $x_0 = 0$ and (8.7). Indeed, for $a \in \ker L$, $a \neq 0$, one has

$$F_0(a) = -tg(a) \int_0^1 ds/p(s) \neq 0$$

and

$$\langle F_0(a), Ja \rangle = -ag(a) \int_0^1 ds/p(s) \leq 0.$$

Notice that $F_0 = QF$, where $F : C_B^1 \to C_0$ is given by

$$F(u)(t) = - \int_0^t \left(\frac{1}{p} g(u) + \frac{p^2 u'}{1 + |pu'|} \right) ds.$$

Finally, we consider the functional $\Phi : C_B^1 \times [0, 1] \to \mathbf{R}$, given by

$$\Phi(u, \lambda) = \frac{1}{2\pi} \left| \int_0^1 \{ pu'^2 + (1 - \lambda) u \left[\frac{1}{p} g(u) + \frac{p^2 u'}{1 + |pu'|} \right] \right.$$

$$+ \lambda pu\, g(u) - \lambda pu f(s, u, pu') \} \, \delta(u)(s) \, ds \left. \right|$$

where $\delta(u)(t) = \min\{1, 1/(u^2 + (pu')^2)\}$. It is easy to see that Φ is continuous.

We have the following existence principle for (8.8)-(8.9).

Theorem 8.8 *Assume (h1) and (h2). Also suppose that*

(h3) *there is an $R \geq 1$ with $\|u\| \leq R$ for each $(u, \lambda) \in \Sigma$ with*

$$\min\nolimits_{[0,1]} \left(u^2 + (pu')^2 \right) < 1;$$

(h4) *for each $n \in \mathbf{N}$ there is $R_n \geq 1$ such that $\|u\| \leq R_n$ for any $(u, \lambda) \in \Sigma$ satisfying*

$$\Phi(u, \lambda) = n \quad and \quad \min\nolimits_{[0,1]} \left(u^2 + (pu')^2 \right) \geq 1.$$

Then (8.8)-(8.9) has at least one solution.

Proof. We apply Corollary 8.7. We first observe that (i1') and (i2') follow from (h3), (h4) since

$$(u, \lambda) \in \Sigma, \ \min_{[0,1]} \left(u^2 + (pu')^2 \right) \geq 1 \implies \Phi(u, \lambda) \in \mathbf{N}. \qquad (8.10)$$

To prove (8.10), let $x_1 = u$, $x_2 = pu'$ and $x = (x_1, x_2)$. Then

$$
\begin{cases}
x_1' = \frac{1}{p}x_2 \\[2mm]
x_2' = -(1 - \lambda)\left[\frac{1}{p} g(x_1) + \frac{px_2}{1+|x_2|} \right] - \lambda pg(x_1) + \lambda pf(t, x_1, x_2)
\end{cases}
\qquad (8.11)
$$

and

$$\Phi(u, \lambda) = \frac{1}{2\pi} \left| \int_0^1 (x_1' x_2 - x_1 x_2') \min\{1, 1/|x|^2\} ds \right|.$$

Consequently, since $|x(t)| \geq 1$ on $[0,1]$, we have

$$\Phi(u, \lambda) = \frac{1}{2\pi} \left| \int_0^1 \left(\arctan \frac{pu'}{u} \right)' ds \right|$$

and so $2\,\Phi(u, \lambda)$ is equal to the (finite) number of (simple) zeroes of u on $[0, 1)$. Here, by a simple zero of u on $[0, 1)$ we mean a number $t \in [0, 1)$ with $u(t) = 0$ and $(pu')(t) \neq 0$.

We now check (i3'). Suppose

$$Lu = (1 - \mu)F_0(u) + \mu H(u, 0)$$

for some $\mu \in (0, 1]$. Then,

$$Lu = (1 - \mu)QF(u) + \mu F(u)$$

or, equivalently,

$$Lu = \mu(F - QF)(u) \quad \text{and} \quad QF(u) = 0.$$

Hence $Lu = \mu F(u)$. It follows that

$$(pu')' = -\mu \left[\frac{1}{p} g(u) + \frac{p^2 u'}{1 + |pu'|} \right]. \qquad (8.12)$$

Multiplying by pu' and letting $G(s) = \int_0^s g(z)dz$, we obtain

$$\left(\frac{1}{2}(pu')^2 + \mu G(u) \right)' = -\mu p \frac{(pu')^2}{1 + |pu'|}. \qquad (8.13)$$

Thus the function $(pu')^2/2 + \mu\, G(u)$ is decreasing. From (8.9), it takes the same value at $t = 0$ and $t = 1$. Consequently, it is constant. Then, from (8.13), it follows that $u' = 0$, that is $u \equiv a$. Now (8.12) guarantees that $0 = -\mu\, g(a)/p$ and so $a = 0$. Also, if for $\mu = 0$ one has $Lu = F_0(u)$, then $u \in \ker L$ and $F_0(u) = 0$, and so $u = 0$. Therefore $\Sigma(0) = \{0\}$ and Corollary 8.7 applies. $\square$

Now we state and prove an auxiliary result due to Capietto, Mawhin and Zanolin [23] which makes it possible to obtain lower bounds for $\Phi(u, \lambda)$ when $(u, \lambda) \in \Sigma$ and $\|u\|$ is large.

Let $f : I \times \mathbf{R}^2 \times [0,1] \to \mathbf{R}^2$ be a Carathéodory function, i.e. $f(\,.\,, y, \lambda)$ is measurable for each (y, λ), $f(t, \,.\,, \,.\,)$ is continuous for a.e. t and, for each $r > 0$, there is $h_r \in L^1(I; \mathbf{R})$ such that $|f(t, y, \lambda)| \le h_r(t)$ for a.e. $t \in I$ and all $|y| \le r$, $\lambda \in [0,1]$. Consider the one-parameter family of problems

$$\begin{cases} y' = f(t, y, \lambda), \\ y(0) = y(1), \end{cases} \tag{8.14}$$

where $\lambda \in [0,1]$. Also define

$$\phi : \left\{ y \in C\left(I; \mathbf{R}^2\right);\ y(0) = y(1) \right\} \times [0,1] \to \mathbf{R}_+,$$

by

$$\phi(y, \lambda) = \frac{1}{2\pi} \left| \int_0^1 [f_1(t, y(t), \lambda)\, y_2(t) - f_2(t, y(t), \lambda)\, y_1(t)]\, z(t)\, dt \right|,$$

where $z(t) = \min\left\{1,\ 1/|y(t)|^2\right\}$. As above, for each solution (y, λ) of (8.14) satisfying $|y(t)| \ge 1$ for all $t \in I$, $2\phi(y, \lambda)$ is the number of zeroes of y_1 on $[0,1)$.

Lemma 8.9 *Suppose that there is a constant* $K_0 > 0$, *a continuous function* $\Theta : \partial B_1(0; \mathbf{R}^2) \to \mathbf{R}_+$ *and a function* $\gamma \in L^1(I; \mathbf{R}_+)$ *such that the following inequality*

$$f_1(t, y, \lambda)\, y_2 - f_2(t, y, \lambda)\, y_1 \ge \alpha_\lambda(t)\, |y|^2\, \Theta(y/|y|) - \gamma(t)\, |y| \tag{8.15}$$

holds for a.e. $t \in I$, *all* $\lambda \in [0,1]$ *and each* $y \in \mathbf{R}^2$ *with* $|y| \ge K_0$, *where for each* $\lambda \in [0,1]$, $\alpha_\lambda \in L^1(I; \mathbf{R})$ *and*

$$\int_0^1 \alpha_\lambda(t)\, dt > 0.$$

Then, for every $\varepsilon > 0$,   there exists a constant $R_\varepsilon \geq 1$ (independent of y and λ) such that, for each couple (y, λ) satisfying (8.14) and $|y(t)| \geq R_\varepsilon$ for   all $t \in I$,   it follows that

$$\phi(y, \lambda) \geq \frac{1}{2\pi \langle \Theta \rangle} \int_0^1 \alpha_\lambda(t)\, dt - \varepsilon,$$

where

$$\langle \Theta \rangle = \frac{1}{2\pi} \int_0^{2\pi} \frac{d\theta}{\Theta(\cos\theta, \sin\theta)}.$$

Proof.   Without loss of generality, we can suppose that $\varepsilon \leq 1$. Let $A \geq \max\{1, K_0\}$ and $\sigma = \min\{\Theta(y); |y| = 1\}$. Suppose that (y, λ) satisfies (8.14) and $|y(t)| \geq A$ for all $t \in I$. Using polar coodinates, we can write

$$y_1(t) = |y(t)| \cos\theta(t), \quad y_2(t) = |y(t)| \sin\theta(t)$$

and

$$\theta'(t) = -[f_1(t, y(t), \lambda)\, y_2(t) - f_2(t, y(t), \lambda)\, y_1(t)] / |y(t)|^2.$$

Then, from (8.15) we obtain

$$\theta'(t) \leq -\alpha_\lambda(t)\, \Theta(\cos\theta(t), \sin\theta(t)) + \gamma(t) / |y(t)|.$$

Consequently

$$\frac{\theta'(t)}{\Theta(\cos\theta(t), \sin\theta(t))} \leq -\alpha_\lambda(t) + \gamma(t) / (\sigma A).$$

Integration over I yields

$$\int_0^1 \frac{\theta'(t)}{\Theta(\cos\theta(t),\, \sin\theta(t))}\, dt = \int_{\theta(0)}^{\theta(1)} \frac{d\theta}{\Theta(\cos\theta,\, \sin\theta)}$$

$$\leq -\|\alpha_\lambda\|_1 + \frac{1}{\sigma A} \|\gamma\|_1.$$

Since $\theta(1) - \theta(0) = 2k\pi$ for some integer k,   if we choose

$$R_\varepsilon = A \geq \|\gamma\|_1 / (2\sigma\pi \langle \Theta \rangle\, \varepsilon),$$

this inequality becomes

$$2k\pi \langle \Theta \rangle \leq - \|\alpha_\lambda\|_1 + 2\pi \langle \Theta \rangle \, \varepsilon.$$

Hence

$$k \leq - \|\alpha_\lambda\|_1 / (2\pi \langle \Theta \rangle) + \varepsilon < \varepsilon \leq 1.$$

Thus $k = -|k|$, and the conclusion follows since $|k| = \phi(y, \lambda)$. $\square$

Returning to our problem, by Lemma 8.9 we can prove the following result.

Lemma 8.10 *For each $n \in \mathbf{N}$ there exists $r_n \geq 1$ such that*

$$\min_{[0,1]} \left(u^2 + (pu')^2 \right) \leq r_n^2 \tag{8.16}$$

for every $(u, \lambda) \in \Sigma$ with $\Phi(u, \lambda) = n$.

Proof. For $y = (y_1, y_2) \in \mathbf{R}^2$, let

$$f_1(t, y, \lambda) = y_2/p \quad \text{and}$$

$$f_2(t, y, \lambda) = - \left[(1 - \lambda) \frac{1}{p} + \lambda p \right] g(y_1)$$

$$- (1 - \lambda) \frac{p y_2}{1 + |y_2|} + \lambda p \, f(t, y_1, y_2).$$

Since $g(y_1)/y_1 \to \infty$ as $|y_1| \to \infty$, for each $j \in \mathbf{N}$, there is a $d_j \geq 0$ with

$$y_1 g(y_1) \geq (2j^2 + c^2/2 + c) y_1^2 - d_j \quad \text{for all} \ y_1 \in \mathbf{R}.$$

Using $1/p \geq p$, we obtain

$$((1 - \lambda)/p + \lambda p) \, y_1 g(y_1) \geq p \left(2j^2 + c^2/2 + c \right) y_1^2 - d_j p^{-1}.$$

On the other hand,

$$-\lambda p \, y_1 f(t, y_1, y_2) \geq -p \, |y_1| \, (c \, |y_1| + c \, |y_2| + k(t))$$

$$\geq -p \left[c y_1^2 + \left(c^2 y_1^2 + y_2^2 \right)/2 + k(t) \, |y_1| \right]$$

and

$$(1 - \lambda) p \, y_1 y_2 / (1 + |y_2|) \geq -p \, |y_1|.$$

For $|y| \geq 1$, these inequalities yield

$$(f_1(t, y, \lambda)y_2 - f_2(t, y, \lambda)y_1) \, / \, |y| \; = \; [y_2^2/p + ((1 - \lambda)/p + \lambda p)y_1 g(y_1)$$

$$+(1 - \lambda)p \, y_1 y_2 \, / (1 + |y_2|) - \lambda p y_1 f(t, y)] \, / \, |y|$$

$$\geq p \, (2j^2 y_1^2 + y_2^2/2) / \, |y| - p(1 + k(t)) - d_j p^{-1}.$$

Thus, inequality (8.15) is fulfilled with

$$\alpha_\lambda(t) \; = \; p(t), \;\; \gamma(t) \; = \; p(t)(1 + k(t)) + d_j p \, (t)^{-1}$$

and

$$\Theta(y) \; = \; 2j^2 y_1^2 + y_2^2 \, /2.$$

Since $\langle \Theta \rangle = 1/j$, from Lemma 8.9, there exists $\bar{r}_j \geq 1$ with

$$\Phi(u, \lambda) \; \geq \; \frac{j}{2\pi} \int_0^1 p(s)ds - 1$$

whenever $(u, \lambda) \in \Sigma$ and $|x(t)| > \bar{r}_j$ for all $t \in [0, 1]$ (recall $x(t) = (u(t), p(t)u'(t))$). Finally, for each $n \in \mathbf{N}$, let $j = j(n)$ be the smallest nonnegative integer satisfying

$$\frac{j}{2\pi} \int_0^1 p(s)ds - 1 > n.$$

Then $r_n = \bar{r}_{j(n)}$ is the number we are looking for. $\square$

Remark 8.3 *Consider the following condition:*

(h5) *for every $r \geq 1$, there is $R(r) \geq 1$ such that $\|u\| \leq R(r)$ for each $(u, \lambda) \in \Sigma$ with $\min_{[0,1]} (u^2 + (pu')^2) \leq r^2$.*

By Lemma 8.10, (h5) clearly implies both (h3) and (h4). For example, (h5) holds for $p \equiv 1$ (see [23]).

In what follows we will give sufficient conditions so that (h3) and (h4) hold.

By a standard extension argument, we shall suppose that all functions on t are 1-periodic and defined on the whole real line. In particular, for each $(u, \lambda) \in \Sigma$ we may suppose that u and pu' are 1-periodic continuous functions on $\mathbf{R}$ and (u, λ) satisfies (8.11) for a.e. $t \in \mathbf{R}$.

Let

$$V(y) = G(y_1) + y_2^2/2 \quad \text{for} \quad y = (y_1, y_2) \in \mathbf{R}^2,$$

where

$$G(y_1) = \int_0^{y_1} g(s)\, ds,$$

and let

$$\rho(r) = \sup \{V(y);\ |y| \le r\} \quad \text{for each}\ \ r \ge 1.$$

Proposition 8.11 *Suppose*

(h6) *there exists* $\beta \in L^1(0, 1; \mathbf{R}_+)$ *and* $\Psi : [0, \infty) \to (0, \infty)$ *continuous, such that*

$$g(y_1)f_1(t, y, \lambda) + y_2 f_2(t, y, \lambda) \le \beta(t)\Psi(V(y)) \tag{8.17}$$

for all $y \in \mathbf{R}^2$, $\lambda \in [0, 1]$ *and a.e.* $t \in [0, 1]$, *and*

$$\int_{\rho(1)}^{\infty} dz\,/\,\Psi(z) > \|\beta\|_1\,. \tag{8.18}$$

Then (h3) is satisfied.

Proof. Let $(u, \lambda) \in \Sigma$ be such that $\min_{[0,1]} |x(t)| < 1$. Let $t_0 \in [0, 1]$ be chosen so that $|x(t_0)| = \min_{[0,1]} |x(t)|$. Then, $V(x(t_0)) \le \rho(1)$. On the other hand, from (8.17),

$$V(x(t))' \le \beta(t)\Psi(V(x(t))) \quad \text{for a.e. } t \in I.$$

Integration from t_0 to t, $t_0 \le t \le t_0 + 1$, yields

$$\int_{\rho(1)}^{V(x(t))} dz\,/\,\Psi(z) \le \int_{V(x(t_0))}^{V(x(t))} dz\,/\,\Psi(z) \le \|\beta\|_1\,,$$

and the conclusion follows by (8.18), since $V(y) \to \infty$ as $|y| \to \infty$. $\quad\square$

Remark 8.4 *If*

$$\int_0^\infty dz \, / \, \Psi(z) \, = \, \infty \,,$$

then inequality (8.18) with $\rho(1)$ replaced by $\rho(r)$, is true for each $r \geq 1$. Note in this case, even (h5) is satisfied. We are in this case for $p \equiv 1$, when (8.17) holds with $\Psi(z) = z + 1$ (see [23]).

Proposition 8.12 *Suppose*

(h2*) $1/p \dot{\in} L^q(0,1)$ *for some* $q > 1$, *and* $k \in L^\infty(0,1)$.

Then (h4) is satisfied.

For the proof we need some lemmas. First we present an inequality of Opial type, essentially due to Beesack and Das [13].

Lemma 8.13 *Assume that $v, w \in C[a,b] \cap C^1(a,b)$, $w > 0$ on (a,b), $1/w \in L^1(a,b)$ and $wv' \in C[a,b]$. If $v(a) = 0$, then*

$$\int_a^b w \, |vv'| \, ds \, \leq \, \sigma \int_a^b wv'^2 ds \,, \tag{8.19}$$

where

$$\sigma \, = \, \left\{ \frac{1}{2} \int_a^b w(t) \left(\int_a^t ds/w(s) \right) dt \right\}^{1/2}.$$

If $v(b) = 0$, then (8.19) holds with

$$\sigma \, = \, \left\{ \frac{1}{2} \int_a^b w(t) \left(\int_t^b ds/w(s) \right) dt \right\}^{1/2}.$$

Proof. Suppose $v(a) = 0$. Let $a < a_1 < b$. Then

$$|v(t)| \, \leq \, |v(a_1)| + \int_{a_1}^t |v'(s)| \, ds, \quad a_1 \leq t < b.$$

By Hölder's inequality, this yields

$$|v(t)| \, \leq \, |v(a_1)| + \left(\int_{a_1}^t ds/w(s) \right)^{1/2} \left(\int_{a_1}^t w(s) \, |v'(s)|^2 \right)^{1/2}$$

for $a_1 \le t < b$. Letting $a_1 \to a$ we obtain

$$|v(t)| \le \left(\int_a^t ds/w(s) \right)^{1/2} \left(\int_a^t w(s) |v'(s)|^2 \right)^{1/2}, \quad a \le t < b. \quad (8.20)$$

Now, set

$$z(t) = \int_a^t w(s) |v'(s)|^2 \, ds. \quad (8.21)$$

Notice

$$z'(t) = w(t) |v'(t)|^2, \quad a < t < b. \quad (8.22)$$

From (8.20), (8.22) and Hölder's inequality, we obtain

$$\int_a^b w \, |v| \, |v'| \, dt = \int_a^b w \, |v| \sqrt{\frac{z'}{w}} \, dt$$

$$\le \int_a^b \left(w(t) \int_a^t ds/w(s) \right)^{1/2} \sqrt{zz'} \, dt$$

$$\le \left\{ \int_a^b w(t) \left(\int_a^t ds/w(s) \right) dt \right\}^{1/2} \left\{ \int_a^b zz' dt \right\}^{1/2}$$

$$= \left\{ \int_a^b w(t) \left(\int_a^t ds/w(s) \right) dt \right\}^{1/2} \left\{ \frac{1}{2} z^2(b) \right\}^{1/2}.$$

This, with (8.21), implies (8.19).The proof is similar when $v(b) = 0$. $\square$

For the next two lemmas, we suppose $(u, \lambda) \in \Sigma$ (also recall the notation $x = (u, pu')$).

Lemma 8.14 *Suppose* $0 < t_2 - t_1 \le 1$, $t_1 \le t_0 \le t_2$, $|x(t_0)| \le r$ *and* $pu' \ge 0$ *(or* ≤ 0) *on* $[t_1, t_2]$. *Then there exists* $\tau \ge 1$ *only depending on* r *(independent of* t_0, t_1, t_2, u *and* λ) *with*

$$|(pu')(t)| \le \tau \quad on \quad [t_1, t_2].$$

Proof. From

$$(pu')' = -(1-\lambda) \frac{p^2 u'}{1 + |pu'|} \quad (8.23)$$

$$- \left[(1-\lambda) \frac{1}{p} + \lambda p \right] g(u) + \lambda p \, f(s, u, pu')$$

if we multiply by pu' and integrate from t_0 to t, we get

$$\frac{1}{2}(pu')^2(t) = \frac{1}{2}(pu')^2(t_0) - (1-\lambda)\int_{t_0}^t p\frac{(pu')^2}{1+|pu'|}\,ds \qquad (8.24)$$

$$- \int_{t_0}^t [(1-\lambda)/p + \lambda p]\,pu'g(u)\,ds + \lambda\int_{t_0}^t p^2u'f(s,u,pu')\,ds.$$

Let $\tau_1 \geq r$ be such that $g(z)/z \geq 1$ for $|z| > \tau_1$. Assume $pu' \geq 0$ on $[t_1, t_2]$ (the case $pu' \leq 0$ on $[t_1, t_2]$ is similar). Let

$$\Upsilon = -(1-\lambda)p\frac{(pu')^2}{1+|pu'|} - \left[(1-\lambda)\frac{1}{p} + \lambda p\right]pu'g(u).$$

First, we estimate the integral $\int_{t_0}^t \Upsilon(s)ds$. We have two cases to consider:

1) Let $t_0 \leq t$. Then $u(t_0) \leq u(s)$ for $t_0 \leq s \leq t_2$.

a) For those s with $u(s) > \tau_1$, we have $-u'g(u) \leq -u'u \leq 0$, and so

$$\int_{[t_0,t]\cap(u>\tau_1)} \Upsilon(s)\,ds \leq 0,$$

where $(u > \tau_1) = \{s;\ u(s) > \tau_1\}$.

b) Also notice

$$\int_{[t_0,t]\cap(|u|\leq\tau_1)} \Upsilon(s)\,ds \leq \|1/p\|_1 \|pu'\|_\infty \max{}_{[-\tau_1,\tau_1]}|g|.$$

c) We have $[t_0, t] \cap (u < -\tau_1) = \emptyset$ because otherwise, from $-r \leq u(t_0) \leq u(s) < -\tau_1$ we can derive $r > \tau_1$, a contradiction. Thus

$$\int_{t_0}^t \Upsilon(s)\,ds \leq \|1/p\|_1 \|pu'\|_\infty \max{}_{[-\tau_1,\tau_1]}|g|.$$

2) Let $t < t_0$. Then $u(s) \leq u(t_0)$ for $t_1 \leq s \leq t_0$.

a) We have $[t, t_0] \cap (u > \tau_1) = \emptyset$ because otherwise, we have $\tau_1 < u(s) \leq u(t_0) \leq r$, a contradiction.

b) Notice

$$-\int_{[t,t_0]\cap(|u|\leq\tau_1)} \Upsilon(s)\,ds \leq \left(1 + \|1/p\|_1 \max{}_{[-\tau_1,\tau_1]}|g|\right)\|pu'\|_\infty.$$

c) Also we have

$$-\int_{[t,t_0]\cap(u<-\tau_1)} \Upsilon(s)\,ds \le \|pu'\|_\infty,$$

because $g(u) \le u < -\tau_1$ and consequently,

$$((1-\lambda)/p + \lambda p)\,pu'g(u) \le 0.$$

Thus

$$\int_{t_0}^t \Upsilon(s)\,ds \le \left(1 + \|1/p\|_1 \max_{[-\tau_1,\tau_1]} |g|\right)\|pu'\|_\infty.$$

Therefore, in both cases, we have

$$\int_{t_0}^t \Upsilon(s)\,ds \le \tau_2 \|pu'\|_\infty, \tag{8.25}$$

where $\tau_2 = 1 + \|1/p\|_1 \max_{[-\tau_1,\tau_1]} |g|$. Notice that (8.25) is also true if $pu' \le 0$ on $[t_1, t_2]$.

Regarding the last term of (8.24) we have

$$\left|\int_{t_0}^t p^2 u' f(s,u,pu')\,ds\right| \le c\left|\int_{t_0}^t \left(p\,|uu'| + (pu')^2\right)ds\right| \tag{8.26}$$

$$+\|k\|_1 \|pu'\|_\infty.$$

From Lemma 8.13, we obtain

$$\left|\int_{t_0}^t p\,|uu'|\,ds\right| \le \left|\int_{t_0}^t p\,|u - u(t_0)|\,|u'|\,ds\right| + r\,\|pu'\|_\infty$$

$$\le \sigma\left|\int_{t_0}^t pu'^2\,ds\right| + r\,\|pu'\|_\infty,$$

where $\sigma = \left(\|p\|_{L^1(I)}\,\|1/p\|_{L^1(I)}\,/\,2\right)^{1/2}$.

Then, from (8.26) we get

$$\left|\int_{t_0}^t p^2 u' f(s,u,pu')\,ds\right| \le (\|k\|_1 + cr)\,\|pu'\|_\infty \tag{8.27}$$

$$+c\,(\sigma + 1)\left|\int_{t_0}^t pu'^2\,ds\right|.$$

For $\delta > 0$ and $1/q + 1/q' = 1$, we have

$$\left| \int_{t_0}^t pu'^2 ds \right| = \left| \int_{t_0}^t (pu')^2 (1/p) ds \right| \le \|1/p\|_q \left| \int_{t_0}^t (pu')^{2q'} ds \right|^{1/q'}$$

$$= \|1/p\|_q \left| \int_{t_0}^t (pu')^{2q'} e^{-2q'\delta|s-t_0|} e^{2q'\delta|s-t_0|} ds \right|^{1/q'}$$

$$\le \|1/p\|_q \left\| pu' e^{-\delta|s-t_0|} \right\|_\infty^2 \left| \int_{t_0}^t e^{2q'\delta|s-t_0|} ds \right|^{1/q'}$$

$$\le \|1/p\|_q \left\| pu' e^{-\delta|s-t_0|} \right\|_\infty^2 (2q'\delta)^{-1/q'} e^{2\delta|t-t_0|}.$$

Now using (8.24), (8.25) and (8.27), we obtain

$$\frac{1}{2}(pu')^2 e^{-2\delta|t-t_0|} \le r^2/2 + (\tau_2 + \|k\|_1 + cr) \left\| pu' e^{-\delta|s-t_0|} \right\|_\infty e^\delta$$

$$+ c(\sigma + 1) \|1/p\|_q (2q'\delta)^{-1/q'} \left\| pu' e^{-\delta|s-t_0|} \right\|_\infty^2. \qquad (8.28)$$

Next choose $\delta > 0$ so that

$$1/2 > c(\sigma + 1) \|1/p\|_q (2q'\delta)^{-1/q'}.$$

Then, from (8.28) it follows that there exists $\tau_3 \ge 1$ with

$$\left| pu' e^{-\delta|t-t_0|} \right| \le \tau_3 \quad \text{on} \quad [t_1, t_2].$$

Hence

$$|(pu')(t)| \le \tau \quad \text{on} \quad [t_1, t_2],$$

where $\tau = \tau_3 e^\delta$. $\square$

Lemma 8.15 *Suppose* $0 < t_2 - t_1 \le 1$, $(pu')(t_1) = (pu')(t_2) = 0$, $pu' \ge 0$ *(or ≤ 0) on* $[t_1, t_2]$, *and* $u > 0$ *(or < 0) on* (t_1, t_2). *Then there exists* $\tau_* \ge 1$ *independent of* t_1, t_2, u *and* λ, *with*

$$|(pu')(t)| \le \tau_* \quad \text{on} \quad [t_1, t_2].$$

Proof. Let $r > 0$ be such that

$$zg(z) > cz^2 + (c + \|k\|_\infty + 1)\,|z| \quad \text{for } |z| > r. \qquad (8.29)$$

We show

$$\text{either } |u(t_1)| \le r \text{ or } |u(t_2)| \le r.$$

Then we apply Lemma 8.14 with $t_0 = t_1$ or $t_0 = t_2$, respectively. Suppose, by contradiction, $|u(t_1)| > r$ and $|u(t_2)| > r$. Then, since u is monotone and preserves its sign on (t_1, t_2), we have $|u(t)| > r$ for all $t \in [t_1, t_2]$. Assume $u(t) > r$ on $[t_1, t_2]$. Then, (8.23), (8.29) with $(1 - \lambda)/p + \lambda p \ge p$ and $g(u(t)) > 0$ establishes

$$(pu')' \le p\,(1 - g(u) + cu + c + \|k\|_\infty) < 0$$

for a.e. t in those neighborhoods of t_1 and t_2 (subintervals of $[t_1, t_2]$) where $|(pu')(t)| \le 1$. Consequently, pu' is strictly decreasing in those neighborhoods of t_1 and t_2. Since $(pu')(t_1) = (pu')(t_2) = 0$, this implies that pu' changes its sign on $[t_1, t_2]$, a contradiction. Repeating the same reasoning for the case $u(t) < -r$ on $[t_1, t_2]$, we complete the proof of Lemma 8.15. □

Lemma 8.16 *Suppose* $0 < t_2 - t_1 \le 1$, $\min_{[t_1, t_2]} |u(t)| \le r$ *and* $|(pu')(t)| \le \tau$ *on* $[t_1, t_2]$. *Then there exists* $r^* \ge r$ *only depending on* r *and* τ, *such that* $|x(t)| \le r^*$ *on* $[t_0, t_1]$.

Proof. Let $t_0 \in [t_1, t_2]$ be such that $|u(t_0)| \le r$. We have $-\tau/p \le u' \le \tau/p$, and so integration from t_0 to t yields

$$-\tau\,\|1/p\|_1 \le u(t) - u(t_0) \le \tau\,\|1/p\|_1.$$

Consequently,

$$|u(t)| \le r + \tau\,\|1/p\|_1 \quad \text{on } [t_1, t_2].$$

Thus, $r^* = \max\{r + \tau\,\|1/p\|_1,\ \tau\}$. □

We are now ready to prove Proposition 8.12.

Proof of Proposition 8.12. Let (u, λ) be any element of Σ satisfying

$$\min_{[0,1]} |x(t)| \ge 1 \quad \text{and} \quad \Phi(u, \lambda) = n.$$

From Lemma 8.10, $\min_{[0,1]} |x(t)| \le r_n$.

Let us consider the nondecreasing sequence s_j, $j \in \mathbf{Z}$, of real numbers with the following properties:

1) $(pu')(s_j) = 0$, $s_{2j} < s_{2j+1}$, $pu' > 0$ (or < 0) on (s_{2j}, s_{2j+1});

2) u changes its sign on (s_{2j}, s_{2j+1}) and preserves the sign on $[s_{2j+1}, s_{2j+2}]$, for all $j \in \mathbf{Z}$.

Since $\Phi(u, \lambda) = n$ and u is 1-periodic on $\mathbf{R}$,

$$s_{j+4nl} = l + s_j \quad \text{for all} \quad j, l \in \mathbf{Z}.$$

Let $j_0 \in \mathbf{Z}$ be such that $\min_{[s_{j_0}, s_{j_0+1}]} |x(t)| \leq r_n$. Two cases are possible:

a) j_0 *is even.* Then, we first use Lemma 8.14 to get $\tau_0 \geq 1$ with

$$|(pu')(t)| \leq \tau_0 \quad \text{on} \quad [s_{j_0}, s_{j_0+1}].$$

Next, from Lemma 8.16, there is an $\underline{r}^0 \geq r_n$ with $|x(t)| \leq \underline{r}^0$ on $[s_{j_0}, s_{j_0+1}]$. Furthermore, we successively obtain $\underline{r}^0 \leq \underline{r}^1 \leq \underline{r}^2 \leq \ldots \leq \underline{r}^{4n-1}$ with

$$|x(t)| \leq \underline{r}^{j-j_0} \quad \text{on} \quad [s_j, s_{j+1}], \quad j_0 \leq j \leq j_0 + 4n - 1.$$

If j is odd, from Lemma 8.15, we have

$$|(pu')(t)| \leq \tau_* \quad \text{on} \quad [s_j, s_{j+1}]$$

and then, from Lemma 8.16 with $t_1 = s_j$, $t_2 = s_{j+1}$, $r = \underline{r}^{j-j_0-1}$ and $\tau = \tau_*$, we get $\underline{r}^{j-j_0}$ as desired. If j is even, we use Lemma 8.14 with $t_1 = t_0 = s_j$, $t_2 = s_{j+1}$ and $r = \underline{r}^{j-j_0-1}$, and we get $\tau_{j-j_0} \geq 1$ with

$$|(pu')(t)| \leq \tau_{j-j_0} \quad \text{on} \quad [s_j, s_{j+1}].$$

Then, from Lemma 8.16 with $t_1 = s_j$, $t_2 = s_{j+1}$, $r = \underline{r}^{j-j_0-1}$ and $\tau = \tau_{j-j_0}$, we find $\underline{r}^{j-j_0} \geq \underline{r}^{j-j_0-1}$ with

$$|x(t)| \leq \underline{r}^{j-j_0} \quad \text{on} \quad [s_j, s_{j+1}].$$

b) j_0 *is odd.* By a similar reasoning, starting this time with Lemma 8.15 instead of Lemma 8.14, we obtain a sequence $\overline{r}^0 \leq \overline{r}^1 \leq \overline{r}^2 \leq \ldots \leq \overline{r}^{4n-1}$ with

$$|x(t)| \leq \overline{r}^{j-j_0} \quad \text{on} \quad [s_j, s_{j+1}], \quad j_0 \leq j \leq j_0 + 4n - 1.$$

Since the length of the interval $[s_{j_0}, s_{j_0+4n}]$ is one,

$$R_n = \max\left\{\underline{r}^{4n-1}, \overline{r}^{4n-1}\right\}$$

is the bound we are looking for. □

Now we can state the following existence result.

Theorem 8.17 *Assume (h1), (h2*) and (h6). Then (8.8)–(8.9) has at least one solution.*

Proof.   Use Propositions 8.11 and 8.12 and apply Theorem 8.8. □

As an example, consider the following problem

$$\frac{1}{p}(pu')' + u^3 = k(t), \quad u(0) = u(1), \quad (pu')(0) = (pu')(1),$$

where $p \in C[0,1] \cap C^1(0,1)$, $0 < p \leq 1$ on $(0,1)$, $1/p \in L^q(0,1)$ for some $q > 1$ and $k \in L^\infty(0,1)$, has at least one solution provided that

$$\left(\pi/2 - \arctan 5/\sqrt{3}\right)/\sqrt{3} > \|\beta\|_1,$$

where $\beta(t) = \max\{1/p(t) - p(t),\ p(t)\,|k(t)|\}$. Indeed, (8.17) holds with $\Psi(z) = 8z^2 + 2z + 1/2$, while $\rho(1) = 1/2$.

For other techniques of "a priori" estimation of solutions based on an analysis of the phase plane we refer the reader to [42] and [89]. For other applications of the general continuation theorems, see [73].

9. The Unified Theory

In this chapter we present some general principles which makes it possible to understand globally particular theorems of Leray-Schauder type for a great variety of single and set-valued maps in metric, locally convex or Banach spaces. In fact, this book could start with this axiomatic theory of the Leray-Schauder type theorems. Nevertheless, we have preferred to introduce successively the main classes of maps for which a continuation theorem is known and immediately give examples of typical applications. After such an excursion the reader could ask himself or herself about the common feature of all theorems of Leray-Schauder type. Some answers can be found in the papers [130]-[132], [134]-[136] and [142]. In what follows we shall give a generalization of the results from [132] and [134].

9.1 A General Continuation Principle

Let Ξ, Θ be two sets and A, B two proper subsets of Ξ and Θ, respectively. Consider a nonempty class of maps

$$M \subset \left\{\Gamma : \Xi \to \Theta; \, \Gamma^{-1}(B) \cap A = \emptyset \right\}$$

whose elements are called *admissible maps* and let

$$\nu : \left\{\Gamma^{-1}(B) \, ; \, \Gamma \in M \right\} \cup \{\emptyset\} \to Z$$

be any map with values in a nonempty set Z.

For each admissible map Γ, the value $\nu\left(\Gamma^{-1}(B)\right)$ stands as a "measure" of the set $\Gamma^{-1}(B)$ of all solutions $x \in \Xi$ to the inclusion $\Gamma(x) \in B$.

Let $\theta = \nu(\emptyset)$. An admissible map Γ is said to be ν-*essential* (*in* M) if

$$\nu\left(\Gamma^{-1}(B)\right) = \nu\left(\Gamma'^{-1}(B)\right) \neq \theta$$

for any admissible map Γ' having the same restriction to A as Γ, i.e. $\Gamma|_A = \Gamma'|_A$. In the opposite case Γ is said to be ν-*inessential* (*in* M). Thus, Γ is ν-inessential if $\nu\left(\Gamma^{-1}(B)\right) = \theta$ or there exists an admissible map Γ' with

$$\Gamma|_A = \Gamma'|_A \quad \text{and} \quad \nu\left(\Gamma^{-1}(B)\right) \neq \nu\left(\Gamma'^{-1}(B)\right).$$

Also consider an equivalence relation $\approx$ on M with

(A) if $\Gamma|_A = \Gamma'|_A$, then $\Gamma \approx \Gamma'$.

We are interested in the case when the equivalence classes contain only ν-essential maps or only ν-inessential maps. A sufficient condition to have such a case is the following one:

(H) if $\Gamma \approx \Gamma'$, then there is an homotopy $\eta : \Xi \times [0,1] \to \Theta$ and a function $\upsilon : \Xi \to [0,1]$ such that

$$\eta(\,.\,,0) = \Gamma, \quad \eta(\,.\,,1) = \Gamma', \quad \eta(\,.\,,\upsilon(\,.\,)) \in M$$

and

$$\upsilon(x) = \begin{cases} 1 & \text{for } x \in \Sigma_\eta \\ 0 & \text{for } x \in A, \end{cases}$$

where $\Sigma_\eta = \{x \in \Xi; \ \eta(x,\lambda) \in B \text{ for some } \lambda \in [0,1]\}$.

Lemma 9.1 *Assume that conditions (A) and (H) hold. Let Γ be an admissible map. Then Γ is ν-inessential if and only if $\nu\left(\Gamma^{-1}(B)\right) = \theta$ or there exists an admissible map Γ' with*

$$\Gamma \approx \Gamma' \quad \text{and} \quad \nu\left(\Gamma^{-1}(B)\right) \neq \nu\left(\Gamma'^{-1}(B)\right). \tag{9.1}$$

Proof. The necessity follows from the definition of the ν-inessential maps and condition (A).

Suppose now that Γ' satisfies (9.1). Let η and υ be the maps associated with Γ and Γ' as in condition (H). If $\Sigma_\eta = \emptyset$, then $\eta(.,0)^{-1}(B) = \emptyset$, i.e. $\Gamma^{-1}(B) = \emptyset$ and so $\nu(\Gamma^{-1}(B)) = \theta$ which means that Γ is ν-inessential. Next, assume $\Sigma_\eta \neq \emptyset$. Let $\Gamma^* = \eta(.,\upsilon(.))$. By (H), $\Gamma^* \in M$. In addition, we have

$$\Gamma|_A = \Gamma^*|_A \quad \text{and} \quad \Gamma'^{-1}(B) = \Gamma^{*-1}(B). \tag{9.2}$$

Therefore

$$\nu\left(\Gamma^{-1}(B)\right) \neq \nu\left(\Gamma'^{-1}(B)\right) = \nu\left(\Gamma^{*-1}(B)\right). \tag{9.3}$$

The relations (9.2) and (9.3) show that Γ is ν-inessential. $\square$

The following result represents the abstract version of the topological transversality theorem.

Theorem 9.2 *Assume that conditions (A) and (H) hold. Let Γ and Γ' be two admissible maps with $\Gamma \approx \Gamma'$. Then Γ and Γ' are both ν-essential or both ν-inessential and in the first case one has*

$$\nu\left(\Gamma^{-1}(B)\right) = \nu\left(\Gamma'^{-1}(B)\right). \tag{9.4}$$

Proof. Suppose that Γ is ν-inessential. If $\nu(\Gamma'^{-1}(B)) = \theta$, then clearly, Γ' is ν-inessential. Thus we may assume that $\nu(\Gamma'^{-1}(B)) \neq \theta$.

In the case when $\nu(\Gamma^{-1}(B)) = \theta$, the ν-inessentiality of Γ' follows from Lemma 9.1 and the symmetry of relation $\approx$. Suppose now that $\nu(\Gamma^{-1}(B)) \neq \theta$ also. Then, Lemma 9.1 implies that there exists an admissible map Γ'' with

$$\Gamma \approx \Gamma'' \quad \text{and} \quad \nu\left(\Gamma^{-1}(B)\right) \neq \nu\left(\Gamma''^{-1}(B)\right).$$

Using the symmetry and the transitivity of relation $\approx$, from $\Gamma \approx \Gamma'$ and $\Gamma \approx \Gamma''$, we obtain $\Gamma' \approx \Gamma''$. Now, if $\nu(\Gamma'^{-1}(B)) \neq \nu(\Gamma''^{-1}(B))$, then Lemma 9.1 applied to Γ' and Γ'' implies that Γ' is ν-inessential, while if $\nu(\Gamma'^{-1}(B)) = \nu(\Gamma''^{-1}(B))$, we deduce that $\nu(\Gamma'^{-1}(B)) \neq \nu(\Gamma^{-1}(B))$ and the ν-inessentiality of Γ' follows once more from Lemma 9.1, this time applied to Γ' and Γ. Therefore, Γ' is also ν-inessential.

Now suppose that Γ and Γ' are ν-essential maps. Clearly, $\nu\left(\Gamma^{-1}(B)\right)$ $\neq \theta$ and $\nu\left(\Gamma'^{-1}(B)\right) \neq \theta$. Since $\Gamma \approx \Gamma'$, we can construct, as in the proof of Lemma 9.1, an admissible map Γ^* with

$$\Gamma|_A = \Gamma^*|_A \quad \text{and} \quad \Gamma'^{-1}(B) = \Gamma^{*-1}(B).$$

Consequently,

$$\nu\left(\Gamma'^{-1}(B)\right) = \nu\left(\Gamma^{*-1}(B)\right). \tag{9.5}$$

On the other hand, since $\Gamma|_A = \Gamma^*|_A$ and Γ is ν-essential, we must have

$$\nu\left(\Gamma^{-1}(B)\right) = \nu\left(\Gamma^{*-1}(B)\right). \tag{9.6}$$

Now (9.4) follows from (9.5) and (9.6). $\square$

There are two kinds of Leray-Schauder type theorems. Firstly, results for large classes of maps for example, completely continuous, set-contractions, condensing or monotone type maps. Theorems 5.4 and 7.2 are of this kind. Secondly, results in terms of a given homotopy. Example of this kind are the continuation theorems for contractions on metric spaces and maps of Mönch type (Theorems 2.4, 5.3 and 7.1). We can understand globally all these results if we look at Theorem 9.2. To see this, let us now suppose that no class of maps is given but only a single map $\eta : \Xi \times [0,1] \to \Theta$ satisfying

$$\eta(x, \lambda) \notin B \quad \text{for all } x \in A, \ \lambda \in [0,1].$$

We choose a family of functions

$$\Delta \subset \{\delta : \Xi \to [0,1]; \ \delta|_A \text{ is constant}\} \tag{9.7}$$

containing the constant functions 0 and 1. Then, we attach to η and Δ the following class of maps

$$M_{\eta,\Delta} = \{\eta(\,.\,,\delta(\,.\,)); \delta \in \Delta\}$$

and define an equivalence relation on $M_{\eta,\Delta}$ by

$$\Gamma \approx \Gamma' \iff$$

$$\Gamma|_A = \Gamma'|_A \text{ or } \{\Gamma|_A, \Gamma'|_A\} = \{\eta_0|_A, \eta_1|_A\}.$$

Here, as usual, $\eta_\lambda = \eta(\,.\,,\lambda)$ for each $\lambda \in [0,1]$. Also consider a function ν defined at least on the following family of sets

$$\left\{\eta(\,.\,,\delta(\,.\,))^{-1}(B); \delta \in \Delta\right\} \cup \{\emptyset\}.$$

Corollary 9.3 *Assume that there exists a function* $v : \Xi \to [0,1]$ *such that:*

$$v(x) = \begin{cases} 1 & \text{for } x \in \Sigma_\eta \\ 0 & \text{for } x \in A; \end{cases}$$

$$(1-v)\delta + v\delta' \in \Delta \quad \text{for all } \delta, \delta' \in \Delta. \tag{9.8}$$

Then η_0 *and* η_1 *are both* v-*essential or both* v-*inessential in* $M_{\eta,\Delta}$ *and*

$$v\left(\eta_0^{-1}(B)\right) = v\left(\eta_1^{-1}(B)\right).$$

Proof. We apply Theorem 9.2. Clearly condition (A) is satisfied. To check (H) let

$$\Gamma, \Gamma' \in M_{\eta,\Delta}, \ \Gamma \approx \Gamma', \ \Gamma = \eta(.,\delta), \ \Gamma' = \eta(.,\delta').$$

Define the homotopy

$$\widehat{\eta}(x,\lambda) = \eta\left(x, (1-\lambda)\delta(x) + \lambda\delta'(x)\right).$$

We have that $\widehat{\eta}(.,0) = \Gamma$ and $\widehat{\eta}(.,1) = \Gamma'$. Then (H) holds for $\widehat{\eta}$ since $\Sigma_{\widehat{\eta}} \subset \Sigma_\eta$. $\square$

Remark 9.1 *Condition (9.8) implies that for each* $\delta \in \Delta$, *there exists* $\delta^* \in \Delta$ *with*

$$\delta^*(x) = \begin{cases} \delta(x) \text{ for } x \in \Sigma_\eta \\ 0 \text{ for } x \in A. \end{cases}$$

(Take $\delta^* = v\delta$*). This is condition (i) in a somewhat more general result [[142], Theorem 2.1].*

Remark 9.2 *In the case when*

$$\eta_{\lambda_1}|_A \not\equiv \eta_{\lambda_2}|_A \quad \text{whenever} \quad \lambda_1 \neq \lambda_2,$$

the above equivalence relation on $M_{\eta,\Delta}$ *can be equivalently defined by*

$$\Gamma \approx \Gamma' \iff \lambda = \lambda' \text{ or } \{\lambda, \lambda'\} = \{0, 1\},$$

where $\Gamma = \eta(.,\delta)$, $\delta|_A \equiv \lambda$, $\Gamma' = \eta(.,\delta')$, $\delta'|_A \equiv \lambda'$. *This particular case was considered in [[134], Theorem 2].*

The existence of a function v like that in Corollary 9.3 is, at least in some particular cases, connected with separation/ extension results of Urysohn type (we refer to [35] for basic facts of general topology). Here are two remarkable versions of Corollary 9.3:

A general continuation principle in normal topological spaces

Corollary 9.4 *Assume that Ξ is a normal topological space, V is a closed set with*

$$\overline{\Sigma}_\eta \subset V \subset \Xi \setminus \overline{A}$$

and that the following condition is satisfied:

$$(1 - v)\,\delta + v\delta' \in \Delta \tag{9.9}$$

for all $\delta,\, \delta' \in \Delta$ and $v \in C\left(\Xi;\, [0,1]\right)$ with $v\left(x\right) = 0$ on $\overline{A}$, $v\left(x\right) = 1$ on V. Then η_0 and η_1 are both ν-essential or both ν-inessential in $M_{\eta,\Delta}$ and

$$\nu\left(\eta_0^{-1}\left(B\right)\right) = \nu\left(\eta_1^{-1}\left(B\right)\right).$$

Proof. Since Ξ is a normal topological space and $\overline{A}$, V are closed disjoint sets, there exists a function $v \in C\left(\Xi;\, [0,1]\right)$ with the required properties. $\square$

Notice that condition (9.9) is satisfied for

$$\Delta = \left\{\delta \in C\left(\Xi;\, [0,1]\right);\ \delta|_A \text{ is constant}\right\}.$$

As we shall see, this is the case for most Leray-Schauder type theorems. However it is not the case for the Leray-Schauder theorem for contractions on metric spaces. Recall that any metric space is normal.

A general continuation principle in completely regular spaces

Corollary 9.5 *Assume that Ξ is a completely regular space, V is a closed set with*

$$\overline{\Sigma}_\eta \subset V \subset \Xi \setminus \overline{A}$$

and in addition that V or $\overline{A}$ is compact. Also assume that (9.9) holds. Then η_0 and η_1 are both ν-essential or both ν-inessential in $M_{\eta,\Delta}$ and

$$\nu\left(\eta_0^{-1}\left(B\right)\right) = \nu\left(\eta_1^{-1}\left(B\right)\right).$$

Proof. Since $\overline{A}$, V are closed sets of a completely regular space and at least one of them is compact, there exists a function v with the desired properties (see [35]). $\square$

Recall that any locally convex space is completely regular.

In Theorem 9.2 no compatibility between the "measure" ν and the class of admissible maps is assumed. In fact, to apply Theorem 9.2 and Corollary 9.3 in practice, we need to identify simple ν-essential maps. This can be achieved by using, for example, methods from degree theory or fixed point theory. For degree theory we refer the reader to [32], [84] and [151].

Throughout this book we have exclusively used as ν the simple indicator function

$$\nu(\Psi) = \begin{cases} 1 & \text{if } \emptyset \neq \Psi \subset \Xi \\ 0 & \text{if } \Psi = \emptyset. \end{cases} \tag{9.10}$$

When ν is given by (9.10), we concisely speak about *essentiality* instead of ν-essentiality.

9.2 A General Fixed Point Continuation Principle

Now we specialize Theorem 9.2 and its corollaries in order to obtain fixed point theorems of Leray-Schauder type. The essentiality of some simple maps will be obtained by using fixed point theorems for self-maps of a certain subset of Ξ.

Consider a set E, Ξ and A two proper subsets of E with $A \subset \Xi$ and $A \neq \Xi$. Also consider a class of maps

$$N \subset \{\gamma : \Xi \to E; \ A \cap \text{Fix } \gamma = \emptyset\}$$

whose elements are called *admissible*. The notation Fix γ stands for the set of all fixed points of γ. An admissible map γ is said to be *essential* if

$$\gamma' \in N \quad \text{and} \quad \gamma'|_A = \gamma|_A \quad \text{implies Fix} \gamma' \neq \emptyset.$$

Let $\sim$ be an equivalence relation on N and assume that the following conditions are satisfied:

(a) if $\gamma|_A = \gamma'|_A$, then $\gamma \sim \gamma'$;

(h) if $\gamma \sim \gamma'$, then there exists $\sigma : \Xi \times [0,1] \to E$ and a function $v : \Xi \to [0,1]$ such that

$$\sigma(.,0) = \gamma, \ \sigma(.,1) = \gamma', \ \sigma(.,v(.)) \in N$$

and

$$v(x) = \begin{cases} 1 & \text{for } x \in \Sigma_\sigma \\ 0 & \text{for } x \in A, \end{cases}$$

where $\Sigma_\sigma = \{x \in \Xi; \ \sigma(x,\lambda) = x \text{ for some } \lambda \in [0,1]\}$.

Theorem 9.6 *Assume that conditions (a) and (h) are satisfied. Let γ and γ' be two admissible maps with $\gamma \sim \gamma'$. Then γ and γ' are both essential or both inessential.*

Proof. Consider $\Theta = E \times E$, $B = \{(x,x); \ x \in E\}$,

$$M = \{\Gamma : \Xi \to \Theta; \ \Gamma(x) = (\gamma(x), x), \text{ where } \gamma \in N\}$$

and define

$$\Gamma \approx \Gamma' \Longleftrightarrow \gamma \sim \gamma'$$

where $\Gamma = (\gamma(.), .)$ and $\Gamma' = (\gamma'(.), .)$. Then Γ is essential (inessential) if and only if the corresponding map γ is. Moreover, (a) is equivalent to (A) and (h) implies (H), if we put $\eta(x,\lambda) = (\sigma(x,\lambda), x)$. Thus we can apply Theorem 9.2. □

If instead of the class N we consider a single homotopy $\sigma : \Xi \times [0,1] \to E$ and a family of functions as in (9.7), we may apply Theorem 9.6 to the class of maps

$$N_{\sigma,\Delta} = \{\sigma(.,\delta(.)); \ \delta \in \Delta\}$$

and to the equivalence relation on $N_{\sigma,\Delta}$

$$\gamma \sim \gamma' \Longleftrightarrow$$

$$\gamma|_A = \gamma'|_A \text{ or } \{\gamma|_A, \gamma'|_A\} = \{\sigma_0|_A, \sigma_1|_A\}.$$

Thus, we have the following fixed point version of Corollary 9.3.

Corollary 9.7 *Assume that there exists a function $v : \Xi \to [0,1]$ satisfying conditions (9.8) with Σ_σ replacing Σ_η. Then σ_0 and σ_1 are both essential or both inessential in $N_{\sigma,\Delta}$.*

The next proposition gives us a scheme to establish the essentiality of some maps in N. It is stated in terms of fixed point structures.

By a *fixed point structure* on a given set E we mean a pair $(P, \mathbf{F})$, where P is a class of nonvoid subsets of E and $\mathbf{F}$ is a map attaching to each set $D \in P$, a family $\mathbf{F}(D)$ of maps from D into D having at least one fixed point each.

Proposition 9.8 *Let $(P, \mathbf{F})$ be a fixed point structure on the set E and let $\gamma_0 \in N$. If for every $\gamma \in N$ satisfying $\gamma|_A = \gamma_0|_A$, there exists $D_\gamma \in P$ and $\tilde{\gamma} \in \mathbf{F}(D_\gamma)$ such that*

$$\tilde{\gamma}|_{\Xi \cap D_\gamma} = \gamma|_{\Xi \cap D_\gamma} \quad and \quad \text{Fix } \tilde{\gamma} \setminus \Xi = \emptyset, \qquad (9.11)$$

then γ_0 is essential.

Proof. Since $\tilde{\gamma} \in \mathbf{F}(D_\gamma)$, there exists an $x \in D_\gamma$ with $\tilde{\gamma}(x) = x$. Then by (9.11), $x \in \Xi$ and so $\gamma(x) = x$. Consequently, γ_0 is essential.
□

Examples

Now we show how the Leray-Schauder type theorems stated in the previous chapters can be derived from the general principles described in this chapter.

1) Theorem 2.4 follows from Corollary 9.7 applied to: $E = K$, $\Xi = \overline{U}$, $\sigma(x, \lambda) = H(x, \lambda)$ and

$$\Delta = \left\{ \delta \in C(\overline{U}; [0, 1]); \ \delta|_{\partial U} \text{ and } \delta|_V \text{ are constant} \right\},$$

where $V = \overline{W}$ and W is the union of open balls $B_r(x)$ when

$$x \in \Sigma_H = \{y; \ H(y, \lambda) = y \text{ for some } \lambda \in [0, 1]\}$$

and $r > 0$ is so small that

$$\Sigma_H \subset W \subset \overline{W} \subset U.$$

In this case, the essentiality of any map in $N_{\sigma, \Delta}$ is equivalent to the equality $\Lambda_H = [0, 1]$, where

$$\Lambda_H = \{\lambda \in [0, 1]; \ H(x, \lambda) = x \text{ for some } x \in U\}.$$

2) Theorem 5.3 follows from Corollary 9.7 and Proposition 9.8 if we put: $E = K$, $\Xi = \overline{U}$, $\sigma(x, \lambda) = (1 - \lambda) x_0 + \lambda T(x)$ and

$$\Delta = \left\{ \delta \in C(\overline{U}; [0, 1]) ; \ \delta|_{\partial U} \text{ is constant} \right\}.$$

Thus, the conclusion of Theorem 5.3 follows from Corollary 9.7 if we prove that the constant map x_0 is essential. For this, let γ be any function in $N_{\sigma, \Delta}$ with $\gamma|_{\partial U} \equiv x_0$. Suppose that

$$\gamma(x) = (1 - \delta(x)) x_0 + \delta(x) T(x), \tag{9.12}$$

where $\delta \in \Delta$. Let

$$D_\gamma = \overline{\text{conv}}\, \gamma(\overline{U})$$

and

$$\tilde{\gamma} : D_\gamma \to D_\gamma, \quad \tilde{\gamma}(x) = \begin{cases} \gamma(x) & \text{for } x \in \overline{U} \\ x_0 & \text{for } x \in D_\gamma \setminus \overline{U}. \end{cases}$$

Let $(P, \mathbf{F})$ be the Mönch's fixed point structure, i.e. P is the class of all nonempty closed convex subsets of X and $\mathbf{F}(D)$ is the set of all continuous maps from D into D satisfying condition (5.1) in Proposition 5.2. Let us observe that $\tilde{\gamma} \in \mathbf{F}(D_\gamma)$. Indeed, let $S \subset D_\gamma$ be countable and

$$\overline{S} = \overline{\text{conv}}\, \{\{x_0\} \cup \tilde{\gamma}(S)\} = \overline{\text{conv}}\, \{\{x_0\} \cup \gamma(S \cap \overline{U})\}.$$

Using the representation formula for γ, (9.12), we see that

$$\overline{\text{conv}}\, \{\{x_0\} \cup \gamma(S \cap \overline{U})\} \subset \overline{\text{conv}}\, \{\{x_0\} \cup T(S \cap \overline{U})\}.$$

Consequently, $\overline{S} \subset \overline{\text{conv}}\, \{\{x_0\} \cup T(S \cap \overline{U})\}$. Since T satisfies (5.2), this yields that $\overline{S} \cap \overline{U}$ is compact. Furthermore, the continuity of T implies that $T(\overline{S} \cap \overline{U})$ is compact and then, by Mazur's lemma, $\overline{\text{conv}}\, \{\{x_0\} \cup T(S \cap \overline{U})\}$ is also compact. It follows that $\overline{S}$ is compact and so $\tilde{\gamma} \in \mathbf{F}(D_\gamma)$. Since (9.11) clearly holds, it follows that the constant map x_0 is essential as claimed.

3) Theorem 5.4 follows from Theorem 9.6 if we take: $E = K$, $\Xi = \overline{U}$, $A = \partial U$, $N = \mathbf{M}$, where $\mathbf{M}$ is $\mathbf{M}_S$, $\mathbf{M}_D$ or $\mathbf{M}_C$, and the equivalence relation on $\mathbf{M}$, $T \sim T'$ if there exists $H : \overline{U} \times [0, 1] \to K$ condensing (set-contraction, compact, respectively) such that H_λ is fixed point free on ∂U for any $\lambda \in [0, 1]$, $T = H_0$ and $T' = H_1$.

4) Theorem 7.2 can be deduced from Theorem 9.2 if we set: $\Xi = \overline{U}$, $A = \partial U$, $\Theta = Y$, $B = \{0\}$,

$$M = \{\Gamma : \overline{U} \to Y; \ \Gamma(x) = Lx - T(x), \ T \in \mathbf{M}^L\}$$

($\mathbf{M}^L$ being $\mathbf{M}_S^L$, $\mathbf{M}_D^L$ or $\mathbf{M}_C^L$), ν is given by (9.10) and $\Gamma \approx \Gamma'$ if there exists an L-condensing (L-set-contraction, L-compact, respectively) map $H : \overline{U} \times [0,1] \to K$ such that $H_0 = L - \Gamma$, $H_1 = L - \Gamma'$ and $H(x, \lambda) \neq Lx$ for all $x \in \partial U$ and $\lambda \in [0,1]$.

An extra example is presented in the next section.

9.3 Continuation Theorems for Maps of Monotone Type

Let $(X, |.|)$ be a Banach space and $\langle ., . \rangle$ be the duality between X^* and X.

A map $T : D \subset X \to X^*$ is said to be

a) *monotone* if

$$\langle T(x) - T(y), x - y \rangle \geq 0 \quad \text{for all } x, y \in D;$$

b) *maximal monotone* if it is monotone and

$$\langle x^* - T(y), x - y \rangle \geq 0 \text{ for all } y \in D \implies x \in D, T(x) = x^*;$$

c) *strongly monotone* if there exists $c > 0$ such that

$$\langle T(x) - T(y), x - y \rangle \geq c|x - y|^2 \quad \text{for all } x, y \in D;$$

d) ψ-*monotone* if there exists a function $\psi : (0, \infty) \to (0, \infty)$ with

$$\limsup_{t \to a} \psi(t) \geq \psi(a) \quad \text{for any } a > 0,$$

such that

$$\langle T(x) - T(y), x - y \rangle \geq \psi(|x - y|) \quad \text{for all } x, y \in D, \ x \neq y.$$

Notice that in the last definition, ψ could be any nondecreasing function from $(0, \infty)$ into $(0, \infty)$.

e) The map T is said to be of *type* (S) if for any sequence $(x_k) \subset D$ with $x_k \to x \in X$ weakly and $\limsup \langle T(x_k), x_k - x \rangle \leq 0$, we have $x_k \to x$ strongly.

Proposition 9.9 *Let T_1 and T_2 two maps from $D \subset X$ into X^*, where D is closed convex. Suppose that T_1 sends bounded sets into relatively compact sets and T_2 is ψ-monotone. Then $T_1 + T_2$ is of type (S).*

Proof. Assume the contrary. Then, there exists a sequence $(x_k) \subset D$ weakly convergent to some $x \in X$, with $|x - x_k| \to a > 0$ and

$$\limsup \langle (T_1 + T_2)(x_k), x_k - x \rangle \leq 0.$$

Clearly, we may assume that $x_k \neq x$ for all k. Since D is closed convex, $x \in D$. Now T_1 sends bounded sets into relatively compact sets, so passing to a subsequence, we may suppose that $(T_1(x_k))$ is strongly convergent and $\langle T_1(x_k), x_k - x \rangle \to 0$. Then

$$\limsup \langle T_2(x_k), x_k - x \rangle \leq 0.$$

On the other hand,

$$\langle T_2(x_k), x_k - x \rangle = \langle T_2(x_k) - T_2(x), x_k - x \rangle + \langle T_2(x), x_k - x \rangle$$

$$\geq \psi(|x - x_k|) + \langle T_2(x), x_k - x \rangle.$$

These with $\psi > 0$ imply that $\psi(|x - x_k|) \to 0$. Also, since $|x - x_k| \to a > 0$, we have

$$\limsup \psi(|x - x_k|) \geq \psi(a).$$

Thus, we obtain $\psi(a) = 0$, a contradiction. □

f) T is said to be *pseudomonotone* if for any sequence $(x_k) \subset D$ with $x_k \to x \in D$ weakly and $\limsup \langle T(x_k), x_k - x \rangle \leq 0$, we have

$$\langle T(x), x - y \rangle \leq \liminf \langle T(x_k), x_k - y \rangle$$

for all $y \in X$.

g) T is *hemicontinuous* if $T(x + ty) \to T(x)$ weakly as $t \to 0$ for every $x \in D$ and $y \in X$.

Proposition 9.10 *Let $T : D \to X^*$ be monotone and hemicontinuous, where $D \subset X$ is open. Then T is pseudomonotone.*

Proof. Let $(x_k) \subset D$ be a sequence with $x_k \to x \in D$ weakly and

$$\limsup \langle T(x_k), x_k - x \rangle \leq 0.$$

Since T is monotone, we have

$$\langle T(x), x_k - x \rangle \leq \langle T(x_k), x_k - x \rangle$$

and so

$$\lim \langle T(x_k), x_k - x \rangle = 0.$$

Let $y \in X$ and $z_t = x + t(y - x)$, $t > 0$. Since D is open and $x \in D$, we have $z_t \in D$ for $0 < t < t_0$. By the monotonicity of T, we have

$$\langle T(x_k) - T(z_t), x_k - z_t \rangle \geq 0.$$

Then

$$\langle T(x_k), x_k - x \rangle - \langle T(z_t), x_k - x + t(x - y) \rangle \geq -t \langle T(x_k), x - y \rangle.$$

Letting $k \to \infty$ and dividing by $-t$, we obtain

$$\langle T(z_t), x - y \rangle \leq \liminf \langle T(x_k), x - y \rangle.$$

Now the hemicontinuity of T implies

$$\langle T(x), x - y \rangle \leq \liminf \langle T(x_k), x - y \rangle.$$

Finally,

$$\liminf \langle T(x_k), x - y \rangle = \liminf (\langle T(x_k), x - x_k \rangle + \langle T(x_k), x_k - y \rangle)$$

$$= \liminf \langle T(x_k), x_k - y \rangle.$$

Thus T is pseudomonotone. $\square$

Notice that any hemicontinuous monotone map $T : X \to X^*$ is maximal monotone (see [119], Corollary III.2.3).

If X is reflexive, then it can be renormed so that X and X^* are both locally uniformly convex. Then the duality map $\mathcal{F}$ is single-valued, bijective, bicontinuous, monotone and of type (S) (see [20], Proposition 8, or [119], III.2.6). In this case, the maximal monotonicity of a monotone map T is equivalent to the surjectivity of $\mathcal{F} + T$. Moreover, if T is maximal monotone, then $\mathcal{F} + T$ is bijective and $(\mathcal{F} + T)^{-1} : X^* \to X$ is demicontinuous (see [119], III.2.11).

For other results on monotone operators we refer the reader to [161].

In what follows we shall state the topological transversality theorem for maps of the form $(\mathcal{F} + T)^{-1}(\mathcal{F} - f)$ with f demicontinuous of type (S) and $T : X \to X^*$ maximal monotone. We shall apply this theorem to establish the existence of solutions to the operator equation $(T + f)(x) = 0$.

Throughout this section X is a reflexive Banach space which is normed so that X and X^* are both locally uniformly convex, $T : X \to X^*$ is a maximal monotone map and $U \subset X$ is a nonempty open bounded subset of X. Let us consider the following class of maps:

$$\mathbf{M} = \{g = (\mathcal{F} + T)^{-1}(\mathcal{F} - f); \ f \text{ demicontinuous of type } (S),$$

$$0 \notin (T + f)(\partial U)\}.$$

A map $g \in \mathbf{M}$ is said to be *essential* in $\mathbf{M}$ if every $g' \in \mathbf{M}$ with $g'(x) = g(x)$ on ∂U has a fixed point.

Theorem 9.11 *Let* $h^* : \overline{U} \times [0,1] \to X^*$ *be such that for every pair of sequences* $(\lambda_k) \subset [0,1]$ *and* $(x_k) \subset \overline{U}$ *with* $\lambda_k \to \lambda$, $x_k \to x$ *weakly and*

$$\limsup \langle h^*(x_k, \lambda_k), x_k - x \rangle \le 0,$$

we have $x_k \to x$ *strongly and* $h^*(x_k, \lambda_k) \to h^*(x, \lambda)$ *weakly. Let*

$$h(x, \lambda) = (\mathcal{F} + T)^{-1}(\mathcal{F}(x) - h^*(x, \lambda))$$

and assume

(i) $h(x, \lambda) \ne x$ *for all* $x \in \partial U$ *and* $\lambda \in [0,1]$;

(ii) h_0 *is essential in* $\mathbf{M}$.

Then h_1 *is essential in* $\mathbf{M}$ *too.*

Proof. The result follows from Theorem 9.6, if we take $\Xi = \overline{U}$, $A = \partial U$, $N = \mathbf{M}$ and $g \sim g'$ if there is a h^* as above with

$$g = (\mathcal{F} + T)^{-1}(\mathcal{F} - h_0^*), \quad g' = (\mathcal{F} + T)^{-1}(\mathcal{F} - h_1^*).$$

First we check condition (a). Suppose g, $g' \in \mathbf{M}$,

$$g = (\mathcal{F} + T)^{-1}(\mathcal{F} - f), \quad g' = (\mathcal{F} + T)^{-1}(\mathcal{F} - f')$$

and $g|_{\partial U} = g'|_{\partial U}$. Then, by the injectivity of $(\mathcal{F} + T)^{-1}$, one has $f|_{\partial U} = f'|_{\partial U}$. Define

$$h^* (.,\lambda) = (1 - \lambda) f + \lambda f'.$$

It is easily seen that h^* satisfies the required condition and hence that $g \sim g'$.

Now we check condition (h). Suppose $g \sim g'$. Let $\Sigma = \{x \in \overline{U};$ $h(x,\lambda) = x$ for some $\lambda \in [0,1]\}$. By (i), $\Sigma \cap \partial U = \emptyset$. In addition Σ is closed. Indeed, let $(x_k) \subset \Sigma$ such that $x_k \to x$ as $k \to \infty$. Since $x_k \in \Sigma$, there is a $\lambda_k \in [0,1]$ with $h(x_k, \lambda_k) = x_k$, or equivalently

$$T(x_k) + h^*(x_k, \lambda_k) = 0.$$

Passing if necessary to a subsequence, we may suppose without loss of generality that $\lambda_k \to \lambda$ and $x_k \to x$ weakly for some $\lambda \in [0,1]$ and $x \in X$. By the monotonicity of T we have

$$\langle h^* (x_k, \lambda_k), x_k - x \rangle = -\langle T(x_k), x_k - x \rangle \leq -\langle T(x), x_k - x \rangle.$$

It follows that $\limsup \langle h^*(x_k, \lambda_k), x_k - x \rangle \leq 0$. Consequently, $x_k \to x$ strongly and $h^*(x_k, \lambda_k) \to h^*(x, \lambda)$ weakly. Now we pass to the limit in

$$\langle -h^*(x_k, \lambda_k) - T(y), x_k - y \rangle = \langle T(x_k) - T(y), x_k - y \rangle \geq 0$$

and we obtain

$$\langle -h^*(x, \lambda) - T(y), x - y \rangle \geq 0.$$

Then, the maximal monotonicity of T, implies $-h^*(x, \lambda) = T(x)$, that is $h(x, \lambda) = x$. This shows that $x \in \Sigma$ and so Σ is closed. Now it is easy to see that condition (h) is satisfied for any function $v \in C(\overline{U}; [0,1])$ with $v(x) = 0$ on ∂U, $v(x) = 1$ on Σ. $\square$

Corollary 9.12 *Let X be a reflexive Banach space which is normed so that X and X^* are locally uniformly convex, $T : X \to X^*$ a maximal monotone map with $T(0) = 0$, $U \subset X$ an open bounded subset with $0 \in U$ and $f_0 : \overline{U} \to X^*$ a demicontinuous map of type (S). Assume that*

$$(1 - \lambda) \mathcal{F}(x) + \lambda f_0(x) + T(x) \neq 0 \quad for \ x \in \partial U, \ \lambda \in [0,1]. \quad (9.13)$$

Then there exists $x \in U$ with

$$(T + f_0)(x) = 0. \quad (9.14)$$

Proof. Equation (9.14) is equivalent to the fixed point problem $x = (\mathcal{F} + T)^{-1}(\mathcal{F} - f_0)(x)$. Observe that

$$0 = (\mathcal{F} + T)^{-1}(\mathcal{F} - \mathcal{F}) \sim (\mathcal{F} + T)^{-1}(\mathcal{F} - f_0),$$

where

$$h^*(\,.\,,\lambda) = (1 - \lambda)\mathcal{F} + \lambda f_0.$$

Hence according to the above theorem, we have only to show that the null map is essential in $\mathbf{M}$.

To do this, let $g \in \mathbf{M}$ be such that $g(x) = 0$ on ∂U. Then $g = (\mathcal{F} + T)^{-1}(\mathcal{F} - f)$, where f is demicontinuous of type (S), $0 \notin (T + f)(\partial U)$ and $f(x) = \mathcal{F}(x)$ on ∂U. We need to show that g has a fixed point, i.e. we must show there exists $x \in U$ with $(T + f)(x) = 0$.

Let X_0 be any finite-dimensional subspace of X and let $P \colon X^* \to X_0^* \equiv X_0$ be the canonical projection of X^* onto X_0^*. Let $U_0 = U \cap X_0$ and

$$G \colon \overline{U}_0 \to X_0, \quad G(x) = P(T + f)(x).$$

Clearly, G is continuous. In addition,

$$\lambda(x - G(x)) \neq x \quad \text{for all } x \in \partial U_0 \text{ and } \lambda \in [0, 1].$$

Indeed, if we suppose that $\lambda(x - G(x)) = x$ for some $x \in \partial U_0$ and $\lambda \in [0, 1]$, then since $x \in \partial U$ and $f(x) = \mathcal{F}(x)$, we have

$$|x|^2 = \lambda \langle x - G(x), x \rangle$$

$$= \lambda\left(|x|^2 - \langle P(T + \mathcal{F})(x), x \rangle\right) = -\lambda \langle T(x), x \rangle.$$

Now since $T(0) = 0$, we have

$$\langle T(x), x \rangle = \langle T(x) - T(0), x - 0 \rangle \geq 0.$$

Thus $x = 0$, which is impossible because $0 \in U$ while $x \in \partial U$. Now, the classical Leray-Schauder principle implies that there exists $x \in U_0$ with $x - G(x) = x$, that is $G(x) = 0$ or equivalently,

$$\langle T(x) + f(x), y \rangle = 0 \quad \text{for all } y \in X_0.$$

Thus the set

$$V_{X_0} = \{x \in U; \ \langle (T + f)(x), x \rangle \leq 0, \ \langle (T + f)(x), y \rangle = 0 \text{ on } X_0\}$$

is nonempty. Clearly, the family $\{V_{X_0}; \ X_0 \subset X, \ \dim X_0 < \infty\}$ has the finite intersection property. Denote by $\overline{V}_{X_0}$ the closure of V_{X_0} with respect to the weak topology on X. Since X is reflexive, $\overline{V}_{X_0}$ is weakly compact and so the family

$$\left\{\overline{V}_{X_0}; \ X_0 \subset X, \ \dim X_0 < \infty\right\}$$

has nonempty intersection. Let x_0 be a point of this intersection. For an arbitrary point $y \in X$ choose X_0 finite-dimensional with $x_0, y \in X_0$. Let $(x_k) \subset V_{X_0}$ be such that $x_k \to x_0$ weakly. Since $x_k \in V_{X_0}$, we have

$$\langle T(x_k) + f(x_k), x_k \rangle \leq 0, \quad \langle T(x_k) + f(x_k), x_0 \rangle = 0, \qquad (9.15)$$

$$\langle T(x_k) + f(x_k), y \rangle = 0.$$

It follows that $\langle (T + f)(x_k), x_k - x_0 \rangle \leq 0$, or equivalently

$$\langle f(x_k), x_k - x_0 \rangle \leq -\langle T(x_k), x_k - x_0 \rangle.$$

Now since

$$\langle T(x_k), x_k - x_0 \rangle \geq \langle T(x_0), x_k - x_0 \rangle$$

and $\langle T(x_0), x_k - x_0 \rangle \to 0$, we deduce that

$$\limsup \langle f(x_k), x_k - x_0 \rangle \leq 0.$$

Since f is of class (S) and demicontinuous, this yields $x_k \to x_0$ strongly and $f(x_k) \to f(x_0)$ weakly. Then, from (9.15), we obtain

$$\langle T(x_k), x_0 \rangle \to -\langle f(x_0), x_0 \rangle \quad \text{and} \quad \langle T(x_k), y \rangle \to -\langle f(x_0), y \rangle.$$

Now,

$$0 \leq \langle T(x_k) - T(y), x_k - y \rangle = \langle T(x_k) - T(y), x_0 - y \rangle$$
$$+ \langle T(x_k), x_k \rangle - \langle T(x_k), x_0 \rangle - \langle T(y), x_k - x_0 \rangle$$
$$\leq \langle T(x_k) - T(y), x_0 - y \rangle - \langle f(x_k), x_k \rangle - \langle T(x_k), x_0 \rangle$$
$$- \langle T(y), x_k - x_0 \rangle,$$

and letting $k \to \infty$ we obtain

$$0 \leq \langle -f(x_0) - T(y), x_0 - y \rangle.$$

Since y is arbitrary in X and T is maximal monotone, we conclude that $-f(x_0) = T(x_0)$ as we wished. $\square$

Next we state a continuation theorem for pseudomonotone maps.

Corollary 9.13 *Let X be a reflexive Banach space which is normed so that X and X^* are locally uniformly convex, $T : X \to X^*$ a maximal monotone map with $T(0) = 0$, $U \subset X$ an open bounded convex subset with $0 \in U$ and $f : \overline{U} \to X^*$ a demicontinuous pseudomonotone map. Assume that*

$$\langle f(x), x \rangle \geq 0 \quad \text{for all } x \in \partial U.$$

Then there exists $x \in \overline{U}$ with $(T + f)(x) = 0$.

Proof. It is easily seen that for each $k \in \mathbf{N}$, $k \geq 1$, the map $f_k = f + \frac{1}{k}\mathcal{F}$ is demicontinuous of type (S) and satisfies $\langle f_k(x), x \rangle > 0$ on ∂U. This immediately yields

$$(1 - \lambda)\mathcal{F}(x) + \lambda f_k(x) + T(x) \neq 0 \quad \text{for } x \in \partial U, \ \lambda \in [0, 1].$$

Thus, from Corollary 9.12, there is an $x_k \in U$ with $T(x_k) + f_k(x_k) = 0$. Hence $T(x_k) + f(x_k) \to 0$. Since X is reflexive and U is bounded and convex, we may suppose that $x_k \to x_0 \in \overline{U}$ weakly. In addition

$$\langle f(x_k), x_k - x_0 \rangle$$

$$= \langle T(x_k) + f(x_k), x_k - x_0 \rangle - \langle T(x_k), x_k - x_0 \rangle$$

$$\leq \langle T(x_k) + f(x_k), x_k - x_0 \rangle - \langle T(x_0), x_k - x_0 \rangle.$$

It follows that

$$\limsup \langle f(x_k), x_k - x_0 \rangle \leq 0.$$

Since f is pseudomonotone, this yields

$$\langle f(x_0), x_0 - y \rangle \leq \liminf \langle f(x_k), x_k - y \rangle.$$

Then

$$\langle -f(x_0) - T(y), x_0 - y \rangle \geq \limsup \langle -f(x_k) - T(y), x_k - y \rangle$$

$$\geq \limsup \langle -f(x_k) - T(x_k), x_k - y \rangle = 0.$$

Consequently, $-f(x_0) = T(x_0)$. $\square$

We note that the above results also follow from the extended degree theory applied to maps of monotone type (see Browder [20]). The elementary approach presented here comes from [132].

For other examples we refer the reader to [131] and [136].

10. Multiplicity

This chapter presents a Leray–Schauder type theory which guarantees the existence of fixed points in shells of Banach spaces. This is turn will enable us to discuss the existence of multiple fixed points for operator equations. In particular we will use the elementary ideas of Chapter 9 to present some fixed point theorems of cone compression or expansion type.

10.1   Leray-Schauder Theorems of Compression-Expansion Type

We feel this chapter should be self contained so for the convenience of the reader we now gather together the results of Chapter 9 that will be needed to discuss multiplicity.

Let K be a closed subset of a Banach space X and U an open subset of K.

1) $M(\overline{U}; K)$ denotes the set of all compact (continuous with relatively compact image) maps $F : \overline{U} \to K$; here $\overline{U}$ denotes the closure of U in K.

2) We let $M_{\partial U}(\overline{U}; K)$ denote the maps $F \in M(\overline{U}; K)$ with $x \neq F(x)$ for $x \in \partial U$; here ∂U denotes the boundary of U in K.

3) A map $F \in M_{\partial U}(\overline{U}; K)$ is *essential* in $M_{\partial U}(\overline{U}; K)$ if for every map $G \in M_{\partial U}(\overline{U}; K)$ with $G|_{\partial U} = F|_{\partial U}$ we have that there exists $x \in \overline{U}$ with $x = G(x)$. Otherwise F is *inessential* in $M_{\partial U}(\overline{U}; K)$ i.e. there exists a map $G \in M_{\partial U}(\overline{U}; K)$ with $G|_{\partial U} = F|_{\partial U}$ and $x \neq G(x)$ for $x \in \overline{U}$.

155

4) Now $F, G \in M_{\partial U}(\overline{U}; K)$ are *homotopic* in $M_{\partial U}(\overline{U}; K)$ written $F \cong G$ in $M_{\partial U}(\overline{U}; K)$ if there exists a compact map $N : \overline{U} \times [0, 1] \to K$ such that $N_\lambda = N(\,.\,, \lambda) : \overline{U} \to K$ belongs to $M_{\partial U}(\overline{U}; K)$ for each $\lambda \in [0, 1]$ and $N_0 = F$, $N_1 = G$.

Theorem 10.1 *Let X, K and U be as above. Suppose F and G are two maps in $M_{\partial U}(\overline{U}; K)$ such that $F \cong G$ in $M_{\partial U}(\overline{U}; K)$. Then F is essential in $M_{\partial U}(\overline{U}; K)$ iff G is essential in $M_{\partial U}(\overline{U}; K)$.*

Throughout this chapter $X = (X, |\,.\,|)$ will be a Banach space and K a closed nonempty subset of X with $\alpha x + \beta y \in K$ for all $\alpha \geq 0$, $\beta \geq 0$ and $x, y \in K$. Let $\rho > 0$ with

$$B_\rho = \{x \in K;\ |x| < \rho\}, \quad S_\rho = \{x \in K;\ |x| = \rho\},$$

and of course $\overline{B}_\rho = B_\rho \cup S_\rho$.

For convenience we recall the following result from Chapter 5.

Theorem 10.2 *Let $X = (X, |\,.\,|)$ be a Banach space and K a closed subset of X with $\alpha x + \beta y \in K$ for all $\alpha \geq 0$, $\beta \geq 0$ and $x, y \in K$. Also let $\rho > 0$ and $v_0 \in K$. Define the mapping $F : \overline{B}_\rho \to K$ by $F(x) = v_0$ for $x \in \overline{B}_\rho$.*
    (i) If $v_0 \in B_\rho$ then F is essential in $M_{S_\rho}(\overline{B}_\rho; K)$.
    (ii) If $v_0 \in K \setminus \overline{B}_\rho$ then F is inessential in $M_{S_\rho}(\overline{B}_\rho; K)$.

Theorem 10.3 *Let $X = (X, |\,.\,|)$ be a Banach space and K a closed nonempty subset of X with $\alpha x + \beta y \in K$ for all $\alpha \geq 0$, $\beta \geq 0$ and $x, y \in K$. Also r, R are constants with $0 < r < R$. Suppose $F \in M(\overline{B}_R; K)$ and assume the following conditions hold:*

$$x \neq F(x) \quad for \quad x \in S_R \cup S_r \tag{10.1}$$

$$\left\{ \begin{array}{l} F : \overline{B}_r \to K \ \ is\ inessential\ in\ M_{S_r}(\overline{B}_r; K) \\ (i.e.\ F|_{\overline{B}_r}\ is\ inessential\ in\ M_{S_r}(\overline{B}_r; K)) \end{array} \right. \tag{10.2}$$

and

$$F : \overline{B}_R \to K \ \ is\ essential\ in\ M_{S_R}(\overline{B}_R; K). \tag{10.3}$$

Then F has a fixed point in $\Omega = \{x \in K;\ r < |x| < R\}$.

Proof. Suppose F has no fixed points in Ω. Now (10.2) implies that there exists a $G \in M(\overline{B}_r; K)$ with $G|_{S_r} = F|_{S_r}$ and with $x \neq G(x)$ for $x \in \overline{B}_r$. Define the map $\Phi : \overline{B}_R \to K$ by

$$\Phi(x) = \begin{cases} F(x), & r < |x| \le R \\ G(x), & |x| \le r. \end{cases}$$

Clearly $\Phi \in M(\overline{B}_R; K)$ and Φ has no fixed points in $\overline{B}_R$ (since G has no fixed points in $\overline{B}_r$ and F has no fixed points in Ω). This contradicts the fact that $F : \overline{B}_R \to K$ is essential in $M_{S_R}(\overline{B}_R; K)$. $\square$

Theorem 10.4 *Let $X = (X, |.|)$ be a Banach space and K a closed nonempty subset of X with $\alpha x + \beta y \in K$ for all $\alpha \ge 0$, $\beta \ge 0$ and $x, y \in K$. Also r, R are constants with $0 < r < R$. Suppose the following conditions are satisfied:*

$$\begin{cases} N : \overline{B}_R \times [0,1] \to K \quad \text{is a compact map} \\ \text{with } N(x,0) = 0 \text{ for all } x \in \overline{B}_R \text{ and such that for} \\ \text{each } \lambda \in [0,1] \quad \text{we have} \quad x \neq N(x,\lambda) \text{ for all } x \in S_R \end{cases} \quad (10.4)$$

$$\begin{cases} H : \overline{B}_r \times [0,1] \to K \quad \text{is a compact map} \\ \text{such that for each } \lambda \in [0,1] \quad \text{we have} \\ x \neq H(x,\lambda) \text{ for all } x \in S_r \end{cases} \quad (10.5)$$

$$H(.,1)|_{\overline{B}_r} = N(.,1)|_{\overline{B}_r} \quad (10.6)$$

and

$$x \neq H(x,0) \quad \text{for all} \quad x \in B_r. \quad (10.7)$$

Then $N(.,1)$ has a fixed point in $\Omega = \{x \in K;\ r < |x| < R\}$.

Proof. We know from Theorem 10.2 that the zero map is essential in $M_{S_R}(\overline{B}_R; K)$. Now (10.4) together with Theorem 10.1 implies

$$N(.,1) : \overline{B}_R \to K \quad \text{is essential in} \quad M_{S_R}(\overline{B}_R; K). \quad (10.8)$$

Also (10.7) (with (10.5)) implies $H(.,0)$ is inessential in $M_{S_r}(\overline{B}_r; K)$. This together with (10.5), (10.6) and Theorem 10.1 implies

$$N(.,1) = H(.,1) : \overline{B}_r \to K \quad \text{is inessential in} \quad M_{S_r}(\overline{B}_r; K). \quad (10.9)$$

Now (10.8), (10.9) (and also (10.4), (10.5) and (10.6)) imply that (10.1), (10.2) and (10.3) of Theorem 10.3 hold with $F(.) = N(.,1)$. The result follows from Theorem 10.3. $\square$

We now present a Leray–Schauder theorem of cone compression type.

Theorem 10.5 *Let* $X = (X, |.|)$ *be a Banach space and* K *a closed nonempty subset of* X *with* $\alpha x + \beta y \in K$ *for all* $\alpha \geq 0$, $\beta \geq 0$ *and* $x, y \in K$. *Also* r, R *are constants with* $0 < r < R$. *Suppose* $F \in M(\overline{B}_R; K)$ *and assume the following conditions hold:*

$$x \neq \lambda F(x) \quad for \quad \lambda \in [0, 1) \quad and \quad x \in S_R \qquad (10.10)$$

and

$$\begin{cases} there\ exists\ a\ \ v \in K \setminus \{0\}\ \ with\ \ x \neq F(x) + \delta v \\ for\ any\ \ \delta > 0\ \ and\ \ x \in S_r. \end{cases} \qquad (10.11)$$

Then F *has a fixed point in* $\{x \in K;\ r \leq |x| \leq R\}$.

Proof. Suppose $x \neq F(x)$ for $x \in S_r \cup S_R$ (otherwise we are finished). Choose $M > 0$ such that

$$|F(x)| \leq M \quad \text{for all} \quad x \in \overline{B}_r.$$

Now choose $\delta_0 > 0$ such that

$$|\delta_0 v| > M + r. \qquad (10.12)$$

Let

$$N(.,\lambda) = \lambda F(.) \quad \text{and} \quad H(.,\lambda) = F(.) + (1 - \lambda) \delta_0 v.$$

Now (10.10) and (10.11) (with $\delta = (1 - \lambda) \delta_0$) imply that (10.4) and (10.5) are satisfied. In addition (10.6) is true since $N(x, 1) = F(x) = H(x, 1)$ for $x \in \overline{B}_r$ and finally (10.12) implies (10.7) is satisfied (note $H(x, 0) = F(x) + \delta_0 v$). The result follows from Theorem 10.4. $\square$

In our next theorem $K \subset X$ will be a cone. Let $\rho > 0$ with

$$\Omega_\rho = \{x \in X;\ |x| < \rho\};\ \ \partial_X \Omega_\rho = \{x \in X;\ |x| = \rho\},$$

$$B_\rho = \{x \in K;\ |x| < \rho\},\ \ S_\rho = \{x \in K;\ |x| = \rho\}.$$

Notice

$$B_\rho = \Omega_\rho \cap K \quad \text{and} \quad S_\rho = \partial_K(\Omega_\rho \cap K) = K \cap \partial_X \Omega_\rho.$$

Theorem 10.6 *Let $X = (X, |.|)$ be a Banach space, $K \subset X$ a cone and let $|.|$ be increasing with respect to K. Also let r, R are constants with $0 < r < R$. Suppose $F : \overline{\Omega}_R \cap K \to K$ is a compact map and assume the following conditions hold:*

$$|F(x)| \leq |x| \quad \text{for all} \quad x \in K \cap \partial_X \Omega_R \qquad (10.13)$$

and

$$|F(x)| > |x| \quad \text{for all} \quad x \in K \cap \partial_X \Omega_r. \qquad (10.14)$$

Then F has a fixed point in $K \cap \{x \in X; \ r \leq |x| \leq R\}$.

Proof. Notice (10.13) implies that (10.10) is true. To see this suppose there exists $x \in S_R$ and $\lambda \in [0, 1)$ with $x = \lambda F(x)$. Then

$$R = |x| = |\lambda| \, |F(x)| < |F(x)| \leq |x| = R,$$

a contradiction. Also (10.14) implies that (10.11) is true. To see this suppose there exists $v \in K \setminus \{0\}$ with $x = F(x) + \delta v$ for some $\delta > 0$ and $x \in S_r$. Now since $|.|$ is increasing with respect to K we have since $\delta v \in K$,

$$|x| = |F(x) + \delta v| \geq |F(x)| > |x|,$$

a contradiction. The result follows from Theorem 10.5. $\square$

For our next two results we will again assume $K \subset X$ is a closed nonempty set with $\alpha x + \beta y \in K$ for all scalars $\alpha \geq 0$, $\beta \geq 0$ and $x, y \in K$.

Theorem 10.7 *Let X be a Banach space and $K \subset X$ a closed nonempty set with $\alpha x + \beta y \in K$ for all $\alpha \geq 0$, $\beta \geq 0$ and $x, y \in K$. Also r, R are constants with $0 < r < R$. Suppose $F \in M(\overline{B}_R; K)$ and assume the following conditions hold:*

$$x \neq F(x) \quad \text{for } x \in S_R \cup S_r \qquad (10.15)$$

$$\begin{cases} F : \overline{B}_r \to K \text{ is essential in } M_{S_r}(\overline{B}_r; K) \\ (i.e. \ F|_{\overline{B}_r} \text{ is essential in } M_{S_r}(\overline{B}_r; K)) \end{cases} \qquad (10.16)$$

and

$$F : \overline{B}_R \to K \text{ is inessential in } M_{S_R}(\overline{B}_R; K). \qquad (10.17)$$

Then F has at least two fixed points x_0 and x_1 with $x_0 \in B_r$ and $x_1 \in \Omega = \{x \in K; \ r < |x| < R\}$.

Proof. We know from (10.16) that F has a fixed point in B_r.

Let $\Psi = F|_{\overline{\Omega}}$ and suppose $\Psi : \overline{\Omega} \to K$ has no fixed points. Now $F : \overline{B}_R \to K$ being inessential in $M_{S_R}(\overline{B}_R; K)$ means that there exists a compact map $G : \overline{B}_R \to K$ with $G|_{S_R} = F|_{S_R}$ and $x \neq G(x)$ for $x \in \overline{B}_R$. Fix $\rho \in (0, r)$ and consider the map Φ given by

$$
\Phi(x) = \begin{cases}
\frac{\rho}{R} G\left(\frac{R}{\rho} x\right), & |x| \leq \rho \\
\frac{(r-\rho)|x|}{(R-\rho)r-(R-r)|x|} \Psi\left(\frac{(R-\rho)r-(R-r)|x|}{(r-\rho)|x|} x\right), & \rho \leq |x| \leq r \\
\Psi(x), & r \leq |x| \leq R.
\end{cases}
$$

Notice $\Phi : \overline{B}_R \to K$ is well defined since if $\rho \leq |x| \leq r$ then

$$
r \leq \left| \frac{(R-\rho)r - (R-r)|x|}{(r-\rho)|x|} x \right| \leq R.
$$

Also $\Phi : \overline{B}_R \to K$ is a compact map. In addition

$$
\Phi|_{S_R} = \Psi|_{S_R} = F|_{S_R} = G|_{S_R}
$$

and

$$
\Phi|_{\overline{\Omega}} = \Psi|_{\overline{\Omega}} = F|_{\overline{\Omega}}
$$

and so Φ has no fixed points in $\overline{B}_R$ (since G has no fixed points in $\overline{B}_R$ and F has no fixed points in $\overline{\Omega}$).

Lets concentrate on $\Phi : \overline{B}_r \to K$ (i.e. $\Phi|_{\overline{B}_r}$). Now

$$
\Phi|_{S_r} = \Psi|_{S_r} = F|_{S_r}
$$

so $\Phi : \overline{B}_r \to K$ is a compact map with $\Phi|_{S_r} = F|_{S_r}$ and Φ has no fixed points in $\overline{B}_r$. This contradicts (10.16). $\square$

Next we present a Leray–Schauder theorem of cone expansion type.

Theorem 10.8 *Let X be a Banach space and $K \subset X$ a closed nonempty set with $\alpha x + \beta y \in K$ for all $\alpha \geq 0$, $\beta \geq 0$ and $x, y \in K$. Also r, R are constants with $0 < r < R$. Suppose $F \in M(\overline{B}_R; K)$ and assume the following conditions hold:*

$$
x \neq \lambda F(x) \quad \text{for } \lambda \in [0, 1) \quad \text{and} \quad x \in S_r \tag{10.18}
$$

and

$$
\begin{cases}
\text{there exists a } v \in K \setminus \{0\} \text{ with } x \neq F(x) + \delta v \\
\text{for any } \delta > 0 \text{ and } x \in S_R.
\end{cases} \tag{10.19}
$$

Then F has a fixed point in $\{x \in K; \ r \leq |x| \leq R\}$.

Proof. Assume $x \neq F(x)$ for $x \in S_r \cup S_R$ (otherwise we are finished). The result follows immediately from Theorem 10.7 once we show conditions (10.16) and (10.17) are satisfied.

Consider the homotopy

$$H : \overline{B}_r \times [0,1] \to K \text{ defined by } H(x,\lambda) = \lambda F(x).$$

Notice $H_0 = 0$, $H_1 = F$ and so since $x \neq F(x)$ on S_r we have $H_0 \cong H_1$ in $M_{S_r}(\overline{B}_r; K)$ (this follows since $H : \overline{B}_r \times [0,1] \to K$ is a compact map with $H_\lambda \in M_{S_r}(\overline{B}_r; K)$ for each $\lambda \in [0,1]$ since (10.18) holds). From Theorem 10.2 we have that $H_0 : \overline{B}_r \to K$ is essential in $M_{S_r}(\overline{B}_r; K)$ and this together with Theorem 10.1 implies $F : \overline{B}_r \to K$ is essential in $M_{S_r}(\overline{B}_r; K)$. Thus (10.16) holds.

Let $\delta_0 > 0$ be such that

$$|\delta_0 v| > \sup_{x \in S_R} |F(x)| + R. \tag{10.20}$$

Consider the homotopy

$$N : \overline{B}_R \times [0,1] \to K \text{ defined by } N(x,\lambda) = F(x) + \lambda \delta_0 v.$$

Notice N is a compact map with $N_0 = F$ and $N_1 = F + \delta_0 v$. Then since (10.19) holds (for all $\delta \geq 0$ since we are assuming $x \neq F(x)$ on S_R), we have

$$N_0 \cong N_1 \text{ in } M_{S_R}(\overline{B}_R; K). \tag{10.21}$$

Notice (10.20) implies for $\lambda \in [0,1]$ and $x \in S_R$ that

$$|\delta_0 v + \lambda F(x)| > R = |x|.$$

Thus

$$x \neq \lambda F(x) + \delta_0 v \text{ if } \lambda \in [0,1] \text{ and } x \in S_R. \tag{10.22}$$

Let $G : \overline{B}_R \to K$ be the constant map $G(x) = \delta_0 v$. Consider the homotopy

$$J : \overline{B}_R \times [0,1] \to K \text{ defined by } J(x,\lambda) = \delta_0 v + \lambda F(x).$$

Now J is a compact map with $J_0 = G$ and $J_1 = N_1$. From (10.22) we have that

$$N_1 \cong G \text{ in } M_{S_R}(\overline{B}_R; K). \tag{10.23}$$

Now (10.21) and (10.23) imply

$$N_0 \cong G \quad \text{in} \quad M_{S_R}(\overline{B}_R; K). \tag{10.24}$$

However since $G(x) = \delta_0\, v$ and $\delta_0\, v \in K \backslash \overline{B}_R$ (see (10.20)) we have from Theorem 10.2 that G is inessential in $M_{S_R}(\overline{B}_R; K)$. Now Theorem 10.1 guarantees that $N_0 = F$ is inessential in $M_{S_R}(\overline{B}_R; K)$. Consequently (10.17) holds. $\square$

Theorem 10.9 *Let* $X = (X, |.|)$ *be a Banach space,* $K \subset X$ *a cone and let* $|.|$ *be increasing with respect to* K. *Also* r, R *are constants with* $0 < r < R$. *Suppose* $F : \overline{\Omega}_R \cap K \to K$ *(here* $\Omega_R = \{x \in X;\ |x| < R\}$*) is a compact map and assume the following conditions hold:*

$$|F(x)| > |x| \quad \text{for all} \quad x \in K \cap \partial_X \Omega_R \tag{10.25}$$

and

$$|F(x)| \le |x| \quad \text{for all} \quad x \in K \cap \partial_X \Omega_r. \tag{10.26}$$

Then F *has a fixed point in* $K \cap \{x \in X;\ r \le |x| \le R\}$.

Proof. Notice (10.25) and (10.26) imply (10.18) and (10.19) are true (see the ideas used in Theorem 10.6). $\square$

We now combine some of the theorems in this chapter (and also Chapter 5) to establish the existence of multiple fixed points.

Theorem 10.10 *Let* $X = (X, |.|)$ *be a Banach space,* $K \subset X$ *a cone and let* $|.|$ *be increasing with respect to* K. *Also* r, R *are constants with* $0 < r < R$. *Suppose* $F : \overline{\Omega}_R \cap K \to K$ *(here* $\Omega_R = \{x \in X;\ |x| < R\}$*) is a compact map and assume the following conditions hold:*

$$x \ne F(x) \quad \text{for all} \quad x \in K \cap \partial_X \Omega_r \tag{10.27}$$

$$|F(x)| > |x| \quad \text{for all} \quad x \in K \cap \partial_X \Omega_R \tag{10.28}$$

and

$$|F(x)| \le |x| \quad \text{for all} \quad x \in K \cap \partial_X \Omega_r. \tag{10.29}$$

Then F *has at least two fixed points* x_0 *and* x_1 *with* $x_0 \in K \cap \Omega_r$ *and* $x_1 \in K \cap (\overline{\Omega}_R \backslash \overline{\Omega}_r)$.

Proof. Now Theorem 5.3 (note (10.29) implies $x \neq \lambda F(x)$ for all $\lambda \in [0,1)$ and $x \in K \cap \partial_X \Omega_r$) implies F has a fixed point $x_0 \in K \cap \overline{\Omega}_r$. In fact (10.27) implies $x_0 \in K \cap \Omega_r$. Also Theorem 10.9 implies F has a fixed point $x_1 \in K \cap \left(\overline{\Omega}_R \setminus \Omega_r\right)$. In fact $x_1 \in K \cap \left(\overline{\Omega}_R \setminus \overline{\Omega}_r\right)$ (To see this note if this is not true then $x_1 \in K$ and $|x_1| = r$, so $x_1 \in K \cap \partial_X \Omega_r$. This contradicts (10.27)). $\square$

Theorem 10.11 *Let* $X = (X, |.|)$ *be a Banach space,* $K \subset X$ *a cone and let* $|.|$ *be increasing with respect to* K. *Also let* L, r, R *be constants with* $0 < L < r < R$. *Suppose* $F : \overline{\Omega}_R \cap K \rightarrow K$ *(here* $\Omega_R = \{x \in X; \ |x| < R\}$*) is a compact map and assume the following conditions hold:*

$$x \neq F(x) \quad \text{for all} \quad x \in K \cap \partial_X \Omega_r \tag{10.30}$$

$$|F(x)| > |x| \quad \text{for all} \quad x \in K \cap \partial_X \Omega_L \tag{10.31}$$

$$|F(x)| \leq |x| \quad \text{for all} \quad x \in K \cap \partial_X \Omega_r \tag{10.32}$$

and

$$|F(x)| > |x| \quad \text{for all} \quad x \in K \cap \partial_X \Omega_R. \tag{10.33}$$

Then F *has at least two fixed points* x_0 *and* x_1 *with* $x_0 \in K \cap (\Omega_r \setminus \Omega_L)$ *and* $x_1 \in K \cap \left(\overline{\Omega}_R \setminus \overline{\Omega}_r\right)$.

Proof. Theorem 10.6 guarantees the existence of $x_0 \in K \cap \left(\overline{\Omega}_r \setminus \Omega_L\right)$. In fact (10.30) implies that $x_0 \in K \cap (\Omega_r \setminus \Omega_L)$. Theorem 10.9 guarantees the existence of x_1. $\square$

10.2 Multiple Solutions of Focal Boundary Value Problems

To show how the theory in this chapter can be applied in practice we next present some results which guarantee the existence of multiple solutions to the $(p, n-p)$ *focal boundary value problem*

$$\begin{cases} (-1)^{n-p} u^{(n)} = \phi(t) f(t, u), & 0 < t < 1 \\ u^{(j)}(0) = 0, & 0 \leq j \leq p-1 \\ u^{(j)}(1) = 0, & p \leq j \leq n-1 \end{cases} \tag{10.34}$$

where $1 \leq p \leq n - 1$ and $n \geq 2$. In particular we will use Theorem 10.10 and Theorem 10.11 to establish the multiplicity results for (10.34). First, however, we gather together some preliminary information that will be needed later. The next result is well known (see [1], [2] for a proof).

Theorem 10.12 *The Green's function $g(t, s)$ for the boundary value problem*

$$\begin{cases} u^{(k)} = 0 \quad on \quad [0, 1] \\ u^{(j)}(t_i) = 0, \quad 0 \leq j \leq m_i - 1 \\ 1 \leq i \leq r, \ r \geq 2, \ 0 \leq t_1 < t_2 < ... < t_r \leq 1, \ \sum_{i=1}^r m_i = k \end{cases}$$

exists and

$$g(t, s) P(t) \geq 0 \quad for \quad (t, s) \in [t_1, t_r] \times [t_1, t_r]$$

where $P(t) = \prod_{i=1}^r (t - t_i)^{m_i}$.

Now suppose $u \in C^{n-1}[0, 1] \cap C^n(0, 1)$ satisfies

$$\begin{cases} (-1)^{n-p} u^{(n)} > 0 \quad on \quad (0, 1) \\ u^{(j)}(0) = 0, \quad 0 \leq j \leq p - 1 \\ u^{(j)}(1) = 0, \quad p \leq j \leq n - 1. \end{cases}$$

We know (e.g. [1], [2]) that for $(t, s) \in [0, 1] \times [0, 1]$,

$$(-1)^{n-p} G^{(j)}(t, s) \geq 0, \quad 0 \leq j \leq p - 1$$

and

$$(-1)^{n-i} G^{(j)}(t, s) \geq 0, \quad p \leq j \leq n - 1;$$

here $G(t, s)$ is the Green's function for

$$\begin{cases} u^{(n)} = 0 \quad on \quad [0, 1] \\ u^{(j)}(0) = 0, \quad 0 \leq j \leq p - 1 \\ u^{(j)}(1) = 0, \quad p \leq j \leq n - 1 \end{cases} \tag{10.35}$$

and $G^{(j)}(t, s) = \frac{\partial^j}{\partial t^j} G(t, s)$. From the above properties of G we have immediately that

$$u^{(i)}(t) \geq 0 \quad for \quad t \in [0, 1], \quad 0 \leq i \leq p$$

and

$$u^{(p+1)}(t) \leq 0 \quad \text{for} \quad t \in [0, 1]$$

with

$$\sup_{t \in [0,1]} |u^{(i)}(t)| = u^{(i)}(1) \quad \text{for} \quad 0 \leq i \leq p - 1.$$

Fix $i \in \{0, 1, ..., p - 1\}$ and let $\phi_i(t) = u^{(i)}(t)$. It is easy to see that $\phi_i(t)$ satisfies the following $p - i + 1$ conditions

$$\phi_i^{(j)}(0) = u^{(i+j)}(0) = 0, \quad j = 0, 1, ..., p - i - 1$$

and

$$\phi_i(1) = u^{(i)}(1);$$

these are conjugate conditions [1]. In addition notice

$$\phi_i^{(p-i+1)}(t) = u^{(p+1)}(t) \leq 0 \quad \text{for} \quad t \in [0, 1].$$

Following [5], $\phi_i(t)$ can be expressed as

$$\phi_i(t) = t^{p-i} \phi_i(1) + \int_0^1 g_1(t, s) \phi_i^{(p-i+1)}(s) \, ds \tag{10.36}$$

where $g_1(t, s)$ is the Green's function for

$$\begin{cases} \phi_i^{(p-i+1)} = 0 \quad \text{on} \quad [0, 1] \\ \phi_i^{(j)}(0) = 0, \quad 0 \leq j \leq p - i - 1 \\ \phi_i(1) = 0. \end{cases}$$

By Theorem 10.12,

$$\text{sgn} \, g_1(t, s) = \text{sgn} \, [t^{p-i}(t - 1)] = -1.$$

Consequently (10.36) implies

$$\phi_i(t) \geq t^{p-i} \phi_i(1) \quad \text{for} \quad t \in [0, 1],$$

i.e.

$$u^{(i)}(t) \geq t^{p-i} u^{(i)}(1) = t^{p-i} \sup_{t \in [0,1]} |u^{(i)}(t)| \quad \text{for} \quad t \in [0, 1].$$

Theorem 10.13 *Suppose* $u \in C^{n-1}[0,1] \cap C^n(0,1)$ *satisfies*

$$\begin{cases} (-1)^{n-p} u^{(n)}(t) > 0 & \text{for } t \in (0,1) \\ u^{(j)}(0) = 0, & 0 \le j \le p-1 \\ u^{(j)}(1) = 0, & p \le j \le n-1. \end{cases}$$

Then

$$u^{(i)}(t) \ge t^{p-i} u^{(i)}(1) = t^{p-i} \sup_{t \in [0,1]} |u^{(i)}(t)| \qquad (10.37)$$

for $t \in [0,1]$ *and* $i \in \{0, 1, ..., p-1\}$.

Also recall the Greens function $G(t, s)$ for (10.35) can be expressed explicitly as

$$G(t,s) = \frac{1}{(n-1)!} \sum_{i=0}^{p-1} \binom{n-1}{i} t^i (-s)^{n-i-1}$$

if $0 \le s \le t$, whereas

$$G(t,s) = -\frac{1}{(n-1)!} \sum_{i=p}^{n-1} \binom{n-1}{i} t^i (-s)^{n-i-1}$$

if $t \le s \le 1$.

We begin by establishing via Theorem 10.10 the existence of twin nonnegative solutions to the $(p, n-p)$ focal problem (10.34).

Theorem 10.14 *Assume the following conditions are satisfied:*

$$\phi \in C(0,1) \quad \text{with } \phi > 0 \quad \text{on } (0,1) \quad \text{and } \phi \in L^1[0,1] \qquad (10.38)$$

$$\begin{cases} f : [0,1] \times [0,\infty) \to [0,\infty) \text{ is continuous with} \\ f(t,u) > 0 \text{ for } (t,u) \in [0,1] \times (0,\infty) \end{cases} \qquad (10.39)$$

$$\begin{cases} f(t,u) \le w(u) \text{ on } [0,1] \times (0,\infty) \text{ with } w \ge 0 \\ \text{continuous and nondecreasing on } [0,\infty) \end{cases} \qquad (10.40)$$

$$\exists\, r > 0 \quad \text{with} \quad \frac{r}{w(r)\, \sup_{t \in [0,1]} \int_0^1 (-1)^{n-p} G(t,s)\, \phi(s)\, ds} > 1 \qquad (10.41)$$

$$\begin{cases} \text{there exists } m \in \left(0, \tfrac{1}{2}\right) \text{ (choose and fix it) and} \\ \tau \in L^1[m,1] \text{ with } \tau > 0 \text{ on } (m,1) \text{ and with} \\ \phi(t) f(t,u) \ge \tau(t)\, w(u) \text{ on } [m,1] \times (0,\infty) \end{cases} \qquad (10.42)$$

and

$$\exists\, R > r \quad with \quad \frac{R}{w\,(m^p\,R)} < \int_m^1 (-1)^{n-p}\, G(\sigma, s)\, \tau(s)\, ds; \qquad (10.43)$$

here $0 \le \sigma \le 1$ *is such that*

$$\int_m^1 (-1)^{n-p}\, G(\sigma, s)\, \tau(s)\, ds = \sup_{t \in [0,1]} \int_m^1 (-1)^{n-p}\, G(t, s)\, \tau(s)\, ds. \quad (10.44)$$

Then (10.34) *has two solutions* $u_1, u_2 \in C^{n-1}[0,1] \cap C^n(0,1)$ *with* $u_1 \ge 0$ *on* $[0,1]$, $u_2 > 0$ *on* $(0,1]$ *and*

$$0 \le \|u_1\|_\infty < r < \|u_2\|_\infty \le R.$$

Proof. To show the existence of u_1, u_2 we will use Theorem 10.10. Let $X = (C[0,1], \| \cdot \|_\infty)$ and

$$K = \{u \in C[0,1] : u \ge 0 \ \text{on} \ [0,1] \ \text{and} \ u(t) \ge t^p \, \|u\|_0 \ \text{for} \ t \in [0,1]\}.$$

Also let $F : K \to C[0,1]$ be defined by

$$F\,(u)\,(t) = \int_0^1 (-1)^{n-p}\, G(t, s)\, \phi(s)\, f(s, y(s))\, ds. \qquad (10.45)$$

A standard argument shows $F : K \to C[0,1]$ is completely continuous. Next we show $F : K \to K$. If $u \in K$, then clearly $F\,(u)\,(t) \ge 0$ for $t \in [0,1]$. Also notice that

$$\begin{cases} (-1)^{n-p}\, F\,(u)^{(n)}\,(t) \ge 0 \ \ \text{on} \ \ (0,1) \\ F\,(u)^{(j)}\,(0) = 0, \ \ 0 \le j \le p-1 \\ F\,(u)^{(j)}\,(1) = 0, \ \ p \le j \le n-1 \end{cases}$$

and so Theorem 10.13 implies

$$F\,(u)\,(t) \ge t^p \, \|F\,(u)\|_\infty \quad \text{for} \ \ t \in [0,1].$$

Consequently $F\,(u) \in K$ so $F : K \to K$. Let

$$\Omega_r = \{u \in C[0,1]; \ \|u\|_\infty < r\} \quad \text{and} \quad \Omega_R = \{u \in C[0,1]; \ \|u\|_\infty < R\}.$$

We first show

$$u \ne F\,(u) \quad \text{for} \ \ y \in K \cap \partial\Omega_r. \qquad (10.46)$$

To see this suppose there exists $u \in K \cap \partial\Omega_r$ with $u = F(u)$. Then $\|u\|_\infty = r$ and also

$$|u(t)| = \left| \int_0^1 (-1)^{n-p} G(t,s) \, \phi(s) \, f(s,u(s)) \, ds \right|$$

$$\leq w(r) \sup_{t \in [0,1]} \int_0^1 (-1)^{n-p} G(t,s) \, \phi(s) \, ds$$

for $t \in [0,1]$. Consequently

$$\frac{r}{w(r) \sup_{t \in [0,1]} \int_0^1 (-1)^{n-p} G(t,s) \, \phi(s) \, ds} \leq 1.$$

This contradicts (10.42) and so (10.46) is true.

Next we show

$$\|F(u)\|_\infty \leq \|u\|_\infty \quad \text{for} \ \ u \in K \cap \partial\Omega_r. \tag{10.47}$$

To see this let $u \in K \cap \partial\Omega_r$. Then $\|u\|_\infty = r$ and for $t \in [0,1]$,

$$F(u)(t) = \int_0^1 (-1)^{n-p} G(t,s) \, \phi(s) \, f(s,u(s)) \, ds$$

$$\leq w(r) \int_0^1 (-1)^{n-p} G(t,s) \, \phi(s) \, ds$$

$$\leq w(r) \sup_{t \in [0,1]} \int_0^1 (-1)^{n-p} G(t,s) \, \phi(s) \, ds < r = \|u\|_\infty.$$

Consequently $\|F(u)\|_\infty \leq \|u\|_\infty$ so (10.47) is true.

Next we show

$$\|F(u)\|_\infty > \|u\|_\infty \quad \text{for} \ \ u \in K \cap \partial\Omega_R. \tag{10.48}$$

To see this let $u \in K \cap \partial\Omega_R$. Then $\|u\|_\infty = R$ and $u(t) \geq t^p R$ for $t \in [0,1]$. In particular $u(t) \geq m^p R$ for $t \in [m,1]$ and so

$$u(t) \in [m^p R, \, R] \quad \text{for} \ \ t \in [m,1].$$

Now with σ as defined in (10.44) we have using (10.42) and (10.43),

$$F(u)(\sigma) = \int_0^1 (-1)^{n-p} G(\sigma,s) \, \phi(s) f(s,u(s)) \, ds$$

$$\geq \int_m^1 (-1)^{n-p} G(\sigma, s) \, \phi(s) f(s, u(s)) \, ds$$

$$\geq \int_m^1 (-1)^{n-p} G(\sigma, s) \, \tau(s) \, w(u(s)) \, ds$$

$$\geq w \, (m^p R) \int_m^1 (-1)^{n-p} G(\sigma, s) \, \tau(s) \, ds > R = \|u\|_\infty .$$

Thus $\|F(u)\|_\infty > \|u\|_\infty$ and so (10.48) is true.

Now Theorem 10.10 implies F has a fixed point $u_2 \in K \cap (\overline{\Omega}_R \setminus \overline{\Omega}_r)$ i.e. $r < \|u_2\|_\infty \leq R$. In addition $u_2(t) \geq t^p \|u_2\|_\infty \geq t^p r$ for $t \in [0, 1]$ so $u_2 > 0$ on $(0, 1]$. Also Theorem 10.10 implies F has a fixed point $u_1 \in K \cap \Omega_r$ i.e. $\|y_1\|_\infty < r$. $\square$

In Theorem 10.14 it is possible for $\|u_1\|_\infty$ to be zero in some applications. Our next theorem guarantees the existence of two solutions $u_1, u_2 \in C^{n-1}[0, 1] \cap C^n(0, 1)$ with $u_1 > 0$ and $u_2 > 0$ on $(0, 1]$.

Theorem 10.15 *Suppose (10.38)-(10.43) hold. In addition assume that there exists L, $0 < L < r$ with*

$$\frac{L}{w \, (m^p L)} < \int_m^1 (-1)^{n-p} G(\sigma, s) \, \tau(s) \, ds \qquad (10.49)$$

Then (10.34) has two solutions $u_1, u_2 \in C^{n-1}[0, 1] \cap C^n(0, 1)$ with $u_1 > 0$, $u_2 > 0$ on $(0, 1]$ and

$$L \leq \|u_1\|_\infty < r < \|u_2\|_\infty \leq R.$$

Proof. We will apply Theorem 10.11. Let X, K and F be as in Theorem 10.14. As in Theorem 10.14 we have $F : K \to K$. Let

$$\Omega_L = \{u \in C[0, 1]; \; \|u\|_\infty < L\} \quad \text{and} \quad \Omega_r = \{u \in C[0, 1]; \; \|u\|_\infty < r\}.$$

As in Theorem 10.14 we have

$$u \neq F(u) \quad \text{and} \quad \|F(u)\|_\infty \leq \|u\|_\infty \quad \text{for} \; u \in K \cap \partial \Omega_r.$$

In addition (follow the reasoning used to prove (10.48) using (10.49) now) we have

$$\|F(u)\|_\infty > \|u\|_\infty \quad \text{for} \; u \in K \cap \partial \Omega_L.$$

Now Theorem 10.11 implies F has a fixed point $u_1 \in K \cap (\Omega_r \setminus \overline{\Omega}_L)$ i.e. $L \leq \|u_1\|_\infty < r$. In addition $u_1(t) \geq t^p \|u_1\|_\infty \geq t^p L$ for $t \in [0,1]$ (so $u_1 > 0$ on $(0,1]$). Also Theorem 10.11 implies F has a fixed point $u_2 \in K \cap (\overline{\Omega}_R \setminus \overline{\Omega}_r)$ i.e. $r < \|u_2\|_\infty \leq R$. Also $u_2(t) \geq t^p r$ for $t \in [0,1]$. $\square$

Example. The boundary value problem

$$\begin{cases} u'' + \frac{1}{2}\left(u^\alpha + u^\beta\right) = 0 \quad \text{on } (0,1) \\ u(0) = u'(1) = 0, \quad 0 < \alpha < 1 < \beta \end{cases} \tag{10.50}$$

has two solutions $u_1,\, u_2 \in C^1[0,1] \cap C^2(0,1)$ with $u_1 > 0,\, u_2 > 0$ on $(0,1]$ and $\|u_1\|_\infty < 1 < \|u_2\|_\infty$.

To see this we will apply Theorem 10.15 with $n = 2$, $p = 1$, $\phi = \tau = 1/2$, $m = 1/4$ and $w(x) = x^\alpha + x^\beta$. Clearly (10.38), (10.39), (10.40) and (10.42 hold. In this situation

$$G(t,s) = \begin{cases} -s, & 0 \leq s \leq t \\ -t, & t \leq s \leq 1 \end{cases}$$

so

$$\sup_{t\in[0,1]} \int_0^1 (-1)^{n-p}\, G(t,s)\, \phi(s)\, ds = \frac{1}{4}.$$

Now (10.41) holds (with $r = 1$) since if $r = 1$,

$$\frac{r}{w(r)\, \sup_{t\in[0,1]} \int_0^1 (-1)^{n-p}\, G(t,s)\, \phi(s)\, ds} = \frac{4r}{r^\alpha + r^\beta} = 2 > 1.$$

Also notice since $\beta > 1$,

$$\frac{x}{w(m\,x)} = \frac{x}{(m\,x)^\alpha + (m\,x)^\beta} \to 0 \quad \text{as } x \to \infty$$

so there exists $R > r = 1$ with (10.43) holding. Finally since $\alpha < 1$,

$$\frac{x}{w(m\,x)} \to 0 \quad \text{as } x \to 0^+$$

so there exists L, $0 < L < r = 1$, with (10.43) holding. Theorem 10.15 now guarantees the result.

Remark 10.1 *It is easy to use the results of Theorem 10.6 and Theorem 10.9 to write criteria which guarantee the existence of more than two fixed points. We leave the details to the reader.*

11. Local Continuation Theorems

All the continuation theorems presented so far were global, even if, in some cases, some local arguments have been used in their proofs.

For a given map $\eta : \mathcal{D} \subset \Xi \times [0,1] \to \Theta$, a subset $B \subset \Theta$ and an element $(x_0, 0) \in \mathcal{D}$ with $\eta(x_0, 0) \in B$, we have tried to prove the existence of a map

$$x : [0,1] \to \Xi \text{ with } x(0) = x_0, \ (x(\lambda), \lambda) \in \mathcal{D} \text{ and}$$

$$\eta(x(\lambda), \lambda) \in B \quad \text{for all } \lambda \in [0,1].$$

Such a function $x(.)$ is called (global) *implicit function* for the inclusion

$$\eta(x, \lambda) \in B \tag{11.1}$$

and the corresponding existence result is said to be a (global) continuation or (global) implicit function theorem. In many cases it is important to know if the implicit function is unique and has some additional properties. For example, when Ξ is a set in a normed space, we can ask about the existence of a continuous or differentiable implicit function. Such properties of the implicit function are reflections of the corresponding properties of η.

Suppose that $(x_0, \lambda_0) \in \mathcal{D}$ is a solution of (11.1). It may happen that (11.1) admits a solution $x(\lambda)$ only for λ in some neighborhood of λ_0. In this case one says that $x(.)$ defines a *local implicit function* and

171

the corresponding existence result is called *local continuation* or *local implicit function theorem.*

It seems to be natural that every global continuation theorem yields a local result. Conversely, we can pass from a local theorem to a global one if we add suitable extra conditions. Thus, we expect that a local theorem will require less hypotheses than a global one. The aim of this short chapter is to state local versions for some of the global continuation theorems presented so far and to suggest their applicability.

In the literature, the local implicit function theorems are most frequently formulated with parameter λ in a suitable topological space Λ. In what follows we shall adopt this usage.

In the first section, we state the local continuation theorem for contractive maps in complete metric spaces and, as a consequence, Robinson's nonsmooth implicit function theorem in Banach spaces. In the next section, the classical smooth implicit function theorem is derived, and in the third section, two special local continuation theorems for completely continuous and ψ-monotone maps are proved. Finally, an application to stable positive solutions of focal boundary value problems is presented.

11.1   Local Continuation Theorems for Contractions

First we state and prove a local implicit function theorem for contractions on complete metric spaces. A special case of this theorem represents one of the main ingredients in the proof of the global continuation Theorem 2.4.

Theorem 11.1 *Let Λ be a topological space, (K, d) a complete metric space and $U \subset K$ open. Let $H : U \times \Lambda \to K$. Suppose that $x_0 \in U$, $\lambda_0 \in \Lambda$ and $H(x_0, \lambda_0) = x_0$. Also assume that $H(x_0, .)$ is continuous at λ_0 and*

$$d\left(H\left(x_1, \lambda\right), H\left(x_2, \lambda\right)\right) \leq \rho\, d\left(x_1, x_2\right) \qquad (11.2)$$

for all $x_1, x_2 \in U$, $\lambda \in \Lambda$, where $\rho < 1$. Then, there exists a neighborhood $V \subset \Lambda$ of λ_0 and a function $x(.)$ from V into U such that $x(\lambda_0) = x_0$ and for each $\lambda \in V$, $x(\lambda)$ is the unique solution in U of the equation $H(x, \lambda) = x$. Also $x(.)$ is continuous at λ_0. Moreover,

1) if for each $x \in U$, $H(x, .)$ is continuous on Λ, then $x(.)$ is continuous on V;

2) if Λ *is a metric space with metric* d_Λ *and there exists* $\rho_1 > 0$
such that

$$d\left(H\left(x, \lambda_1\right), H\left(x, \lambda_2\right)\right) \leq \rho_1 d_\Lambda\left(\lambda_1, \lambda_2\right) \qquad (11.3)$$

for all $x \in U$ *and* $\lambda_1, \lambda_2 \in \Lambda$, *then* $x\left(.\right)$ *satisfies*

$$d\left(x\left(\lambda_1\right), x\left(\lambda_2\right)\right) \leq \frac{\rho_1}{1-\rho} d_\Lambda\left(\lambda_1, \lambda_2\right), \quad \text{for all } \lambda_1, \lambda_2 \in V; \quad (11.4)$$

3) if $\Lambda \subset \mathbf{R}$, *then there exists a maximal interval* $V \subset \Lambda$ *with these properties, and under the assumption of 1), this maximal interval* V *is open in* Λ.

Proof. Let $\delta > 0$ be such that $\overline{B}_\delta\left(x_0\right) \subset U$. By the continuity of $H\left(x_0, .\right)$ at λ_0, there exists a neighborhood V of λ_0 with

$$d\left(x_0, H\left(x_0, \lambda\right)\right) \leq \left(1 - \rho\right) \delta$$

for $\lambda \in V$. Then

$$d\left(x_0, H\left(x, \lambda\right)\right) \leq d\left(x_0, H\left(x_0, \lambda\right)\right) + d\left(H\left(x_0, \lambda\right), H\left(x, \lambda\right)\right)$$

$$\leq \left(1 - \rho\right) \delta + \rho\, d\left(x_0, x\right) \leq \delta$$

for all $x \in \overline{B}_\delta\left(x_0\right)$ and $\lambda \in V$. Hence, for each $\lambda \in V$, $H_\lambda \doteq H\left(., \lambda\right)$ is a self-map of $\overline{B}_\delta\left(x_0\right)$. From Banach's contraction principle, it has a unique fixed point $x\left(\lambda\right)$.
    We have

$$d\left(x\left(\lambda_1\right), x\left(\lambda_2\right)\right) = d\left(H\left(x\left(\lambda_1\right), \lambda_1\right), H\left(x\left(\lambda_2\right), \lambda_2\right)\right)$$

$$\leq d\left(H\left(x\left(\lambda_1\right), \lambda_1\right), H\left(x\left(\lambda_2\right), \lambda_1\right)\right) + d\left(H\left(x\left(\lambda_2\right), \lambda_1\right), H\left(x\left(\lambda_2\right), \lambda_2\right)\right)$$

$$\leq \rho\, d\left(x\left(\lambda_1\right), x\left(\lambda_2\right)\right) + d\left(H\left(x\left(\lambda_2\right), \lambda_1\right), H\left(x\left(\lambda_2\right), \lambda_2\right)\right).$$

Hence

$$d\left(x\left(\lambda_1\right), x\left(\lambda_2\right)\right) \leq \frac{1}{1-\rho} d\left(H\left(x\left(\lambda_2\right), \lambda_1\right), H\left(x\left(\lambda_2\right), \lambda_2\right)\right). \quad (11.5)$$

From this, under the assumption of 1), it follows that $x\left(.\right)$ is continuous at each $\lambda_2 \in V$.
    When we only have that $H\left(x_0, .\right)$ is continuous at λ_0, then, if in (11.5) we set $\lambda_2 = \lambda_0$, one finds that $x\left(.\right)$ is continuous at λ_0.

Furthermore, (11.4) follows from (11.5) and (11.3).

Now, assume that the conditions of 3) hold. If $(V_1, x_1(\,.\,))$ and $(V_2, x_2(\,.\,))$ are two couples with the required properties, then from (11.2), one deduces that $x_1(\lambda) = x_2(\lambda)$ for all $\lambda \in V_1 \cap V_2$ and so, the maximal interval V is the union of all such intervals. It is open in Λ because of its maximality. $\square$

The following local implicit function theorem due to Robinson [148] is the variant of Theorem 11.1 for equations of the form $F(x, \lambda) = 0$.

Theorem 11.2 *Let* (K, d) *be a complete metric space,* Λ *a topological space and* Y *a normed linear space. Let* $x_0 \in K$, $\lambda_0 \in \Lambda$, U_1 *an open neighborhood of* x_0 *and* $\lambda_0 \in \Lambda$. *Suppose that* $F : U_1 \times \Lambda \to Y$, $F(x_0, \lambda_0) = 0$, *and there exists an one-to-one map* $A : U_1 \to Y$ *such that the following conditions are satisfied:*

(a) A *strongly approximates* F *in* x *at* (x_0, λ_0), *i.e. for each* $\varepsilon > 0$, *there exist neighborhoods* U_ε *and* V_ε *of* x_0 *and* λ_0 *such that*

$$|F(x_1, \lambda) - F(x_2, \lambda) - A(x_1) + A(x_2)| \leq \varepsilon\, d(x_1, x_2)$$

for all $x_1, x_2 \in U_\varepsilon$ *and* $\lambda \in V_\varepsilon$;

(b) $F(x_0, \,.\,)$ *is continuous at* λ_0;

(c) $A(U_1)$ *is a neighborhood of* $A(x_0)$ *in* Y;

(d) A^{-1} *is Lipschitzian with Lipschitz constant* $l_0 > 0$.

Then, there exist neighborhoods $U \subset U_1$ *and* $V \subset \Lambda$ *of* x_0 *and* λ_0 *respectively, and a function* $x : V \to U$ *such that* $x(\lambda_0) = x_0$ *and for each* $\lambda \in V$, $x(\lambda)$ *is the unique solution in* U *of* $F(x, \lambda) = 0$. *Also* $x(\,.\,)$ *is continuous at* λ_0. *Moreover,*

1) if for each $x \in U_1$, $F(x, \,.\,)$ *is continuous on* Λ, *then* $x(\,.\,)$ *is continuous on* V;

2) if Λ *is a metric space with metric* d_Λ *and for each* $x \in U_1$, $F(x, \,.\,)$ *is Lipschitzian on* Λ *with Lipschitz constant* l_1, *then for each* $l > l_0 l_1$, *there is a set* V_l *as above so that* $x(\,.\,)$ *is Lipschitzian on* V_l *with Lipschitz constant* l.

Proof. From (c), there is an $r > 0$ with $B_r(A(x_0)) \subset A(U_1)$. Choose $\varepsilon > 0$ with $\varepsilon l_0 < 1$. From (a) it follows that there is an $\alpha > 0$ and a neighborhood $V_0 \subset \Lambda$ of λ_0 such that

$$|F(x_1, \lambda) - F(x_2, \lambda) - A(x_1) + A(x_2)| \leq \varepsilon\, d(x_1, x_2)$$

for all $x_1, x_2 \in B_\alpha(x_0) \subset U_1$ and $\lambda \in V_0$. On the other hand, since $F(x_0, \lambda_0) = 0$, for all $x \in B_\alpha(x_0)$ and $\lambda \in V_0$, we have

$$|A(x) - F(x, \lambda) - A(x_0)|$$

$$\leq |F(x, \lambda) - F(x_0, \lambda) - A(x) + A(x_0)| + |F(x_0, \lambda)|$$

$$\leq \varepsilon\, d(x, x_0) + |F(x_0, \lambda) - F(x_0, \lambda_0)|$$

$$< \varepsilon\, \alpha + |F(x_0, \lambda) - F(x_0, \lambda_0)|.$$

Taking into account the continuity of $F(x_0, .)$ at λ_0, we may assume that α and V_0 are small enough that

$$\varepsilon\, \alpha + |F(x_0, \lambda) - F(x_0, \lambda_0)| \leq r.$$

Then $|A(x) - F(x, \lambda) - A(x_0)| < r$. Hence

$$A(x) - F(x, \lambda) \in B_r(A(x_0); Y) \subset A(U_1)$$

for all $x \in U := B_\alpha(x_0; K)$ and $\lambda \in V_0$. Consequently, the map $H(x, \lambda) = A^{-1}(A(x) - F(x, \lambda))$ is well-defined from $U \times V_0$ into K. We now check that (11.2) holds for $\rho = \varepsilon l_0$. Indeed, for every $x_1, x_2 \in U$ and $\lambda \in V_0$, we have

$$d(H(x_1, \lambda), H(x_2, \lambda)) \leq l_0 |F(x_1, \lambda) - F(x_2, \lambda) - A(x_1) + A(x_2)|$$
$$\leq \varepsilon l_0\, d(x_1, x_2).$$

All the other assumptions are fulfilled, so Theorem 11.1 applies.

Assume that the additional assumption of 2) holds. Then, since $l > l_0 l_1$, ε can be chosen such that $l = l_0 l_1 / (1 - \varepsilon l_0)$. Now the conclusion follows from Theorem 11.1. 2), where $\rho = \varepsilon l_0$ and $\rho_1 = l_0 l_1$. $\square$

11.2   The Classical Implicit Function Theorem

We shall derive the classical implicit function theorem from Theorem 11.2. We first introduce some notions and notation.

Let X and Y be two real Banach spaces and let $\mathbf{L}(X, Y)$ be the space of all linear continuous maps from X into Y. Let $U \subset X$ be open. A map $F : U \to Y$ is said to be (*Fréchet-*) *differentiable* at $x_0 \in U$ if there exists an $F'(x_0) \in \mathbf{L}(X, Y)$ such that

$$F(x) - F(x_0) = F'(x_0)(x - x_0) + \omega(x_0, x)$$

for all $x \in U$, where $\omega(x_0, x) = o(|x - x_0|)$, i.e.

$$\omega(x_0, x) / |x - x_0| \to 0 \quad \text{as } x \to x_0.$$

In this case $F'(x_0)$ is called the (*Fréchet-*) *derivative* of F at x_0. If F is differentiable at each $x \in U$ and $F' : U \to \mathbf{L}(X, Y)$ is continuous, then F is said to be *continuously differentiable* on U and we denote this by $F \in C^1(U; Y)$.

In case that F depends on two variables x and λ, we use the notations F'_x and F'_λ for the corresponding partial derivatives.

Theorem 11.3 *Let X, Λ, Y be Banach spaces, $U_1 \subset X$ and $V_1 \subset \Lambda$ open neighborhoods of x_0 and λ_0, respectively. Suppose that*

   (i) $F : U_1 \times V_1 \to Y$ and $F(x_0, \lambda_0) = 0$;

   (ii) $F(x_0, .)$ is continuous at λ_0;

   (iii) there exists F'_x on $U_1 \times V_1$, F'_x is continuous at (x_0, λ_0) and the map $F'_x(x_0, \lambda_0) \in \mathbf{L}(X, Y)$ is invertible.

   Then, there exist neighborhoods $U \subset U_1$ and $V \subset V_1$ of x_0 and λ_0 respectively, and a function $x : V \to U$ such that $x(\lambda_0) = x_0$ and for each $\lambda \in V$, $x(\lambda)$ is the unique solution of $F(x, \lambda) = 0$. Also, $x(.)$ is continuous at λ_0. Moreover,

   1) if for each $x \in U_1$, $F(x, .)$ is continuous on V_1, then $x(.)$ is continuous on V;

   2) if for each $x \in U_1$, $F(x, .)$ is Lipschitzian on V_1 with Lipschitz constant l_1, then for each $l > l_0 l_1$, where $l_0 = \left\| [F'_x(x_0, y_0)]^{-1} \right\|$, there is such a set V_l so that $x(.)$ is Lipschitzian on V_l with Lipschitz constant l.

Proof.   We apply Theorem 11.2 to $A = F'_x(x_0, \lambda_0)$. Since $F'_x(x_0, \lambda_0)$ $\in \mathbf{L}(X, Y)$, by the open mapping theorem, $A(U_1)$ is open and A^{-1}

$\in \mathbf{L}(Y, X)$. Thus conditions (c) and (d) of Theorem 11.2 are satisfied. We now check condition (a). For every $x_1, x_2 \in U_1$ and $\lambda \in V_1$, let

$$f(t) = F(tx_1 + (1 - t)x_2, \lambda), \quad t \in [0, 1].$$

Then

$$F(x_1, \lambda) - F(x_2, \lambda) = f(1) - f(0) = \int_0^1 f'(t)\, dt$$

$$= \int_0^1 F'_x(tx_1 + (1 - t)x_2, \lambda)(x_1 - x_2)\, dt.$$

Hence

$$F(x_1, \lambda) - F(x_2, \lambda) - A(x_1 - x_2)$$

$$= \int_0^1 (F'_x(tx_1 + (1 - t)x_2, \lambda) - A)(x_1 - x_2)\, dt.$$

Consequently

$$|F(x_1, \lambda) - F(x_2, \lambda) - A(x_1 - x_2)|$$

$$\leq |x_1 - x_2| \int_0^1 \|F'_x(tx_1 + (1 - t)x_2, \lambda) - A\|\, dt.$$

Let $\varepsilon > 0$. Since F'_x is continuous at (x_0, λ_0), there is a $\delta > 0$ with

$$\|F'_x(x, \lambda) - F'_x(x_0, \lambda_0)\| = \|F'_x(x, \lambda) - A\| \leq \varepsilon$$

whenever $|x - x_0| < \delta$ and $|\lambda - \lambda_0| < \delta$. Then, if $x_1, x_2 \in U_\varepsilon = B_\delta(x_0; X)$ and $\lambda \in V_\varepsilon = B_\delta(\lambda_0; \Lambda)$, we also have $tx_1 + (1 - t)x_2 \in B_\delta(x_0; X)$ and so

$$\|F'_x(tx_1 + (1 - t)x_2, \lambda) - A\| \leq \varepsilon.$$

Therefore

$$|F(x_1, \lambda) - F(x_2, \lambda) - A(x_1 - x_2)| \leq \varepsilon |x_1 - x_2|.$$

Thus, Theorem 11.2 applies. $\square$

Proposition 11.4 *Suppose that all the assumptions of Theorem 11.3 are satisfied. In addition, assume that there exists F'_λ on $U_1 \times V_1$ and F'_λ is continuous at (x_0, λ_0). Then the implicit function $x(.)$ is differentiable at λ_0 and*

$$x'(\lambda_0) = -[F'_x(x_0, \lambda_0)]^{-1} F'_\lambda(x_0, \lambda_0).$$

Proof. Without loss of generality, we may suppose that $x_0 = 0$ and $\lambda_0 = 0$. We have to show that for each $\varepsilon > 0$, there exists $\delta > 0$ such that for all $\lambda \in V$ with $|\lambda| < \delta$, we have

$$|x(\lambda) - x(0) - B(\lambda)| = |x(\lambda) - B(\lambda)| \le \varepsilon |\lambda|,$$

where

$$B = -\left[F_x'(0,0)\right]^{-1} F_\lambda'(0,0).$$

Indeed, we have

$$x(\lambda) - B\lambda = A^{-1}Ax(\lambda) + A^{-1}F_\lambda'(0,0)\lambda$$

$$= A^{-1}\left(F_x'(0,0)x(\lambda) + F_\lambda'(0,0)\lambda\right).$$

Since $F(x(\lambda), \lambda) = F(0,0) = 0$, this yields

$$|x(\lambda) - B\lambda|$$

$$\le \left\|A^{-1}\right\| \left|F(x(\lambda), \lambda) - F(0,0) - F_x'(0,0)x(\lambda) - F_\lambda'(0,0)\lambda\right|.$$

On the other hand, since F_x', F_λ' exist in a neighborhood of the origin and are continuous at origin, we have

$$\left|F(x, \lambda) - F(0,0) - F_x'(0,0)x - F_\lambda'(0,0)\lambda\right|$$

$$= \left|\int_0^1 F_\lambda'(x, t\lambda)\lambda\, dt + \int_0^1 F_x'(tx, 0)x\, dt - F_x'(0,0)x - F_\lambda'(0,0)\lambda\right|$$

$$\le \int_0^1 \left[\left\|F_x'(tx, 0) - F_x'(0,0)\right\| |x| + \left\|F_\lambda'(x, t\lambda) - F_\lambda'(0,0)\right\| |\lambda|\right] dt$$

$$\le \eta \left[|x| + |\lambda|\right]$$

provided that (x, λ) is sufficiently close to the origin. Hence

$$|x(\lambda) - B(\lambda)| \le \eta \left[|x(\lambda)| + |\lambda|\right]$$

$$\le \eta \left[|x(\lambda) - B(\lambda)| + |B(\lambda)| + |\lambda|\right].$$

Then, if we take $\eta < 1$, we have

$$|x(\lambda) - B(\lambda)| \le \eta |\lambda| (1 + \|B\|) / (1 - \eta).$$

Finally, taking η such that $\eta (1 + \|B\|) / (1 - \eta) \le \varepsilon$, we obtain that $|x(\lambda) - B(\lambda)| \le \varepsilon |\lambda|$ for λ close enough to $\lambda_0 = 0$. $\square$

Corollary 11.5 *Let X and Y be Banach spaces, $U_1 \subset X$ open, $V_1 \subset \mathbf{R}$, $x_0 \in U_1$ and $\lambda_0 \in V_1$ interior to V_1. Suppose that*

$$F \in C^1 (U_1 \times V_1; Y), \quad F(x_0, \lambda_0) = 0$$

and

$$F'_x (x_0, \lambda_0) \in \mathbf{L}(X, Y) \quad \text{is invertible.}$$

Then, there exists a maximal interval $V \subset V_1$ with $\lambda_0 \in V$ and a function $x(.) \in C^1(V; U_1)$ such that $x(\lambda_0) = x_0$, and for each $\lambda \in V$,

$$F(x(\lambda), \lambda) = 0 \quad \text{and} \quad F'_x (x(\lambda), \lambda) \quad \text{is invertible.}$$

In addition, this maximal interval is open in V_1.

Proof. The existence of the maximal interval with the desired properties is a consequence of the local uniqueness of the solutions of the equations $F(., \lambda) = 0$. The property of V being open in V_1 is the consequence of the local continuation possible in a neighborhood of each zero (x, λ) of F with λ interior to V_1 and $F'_x (x, \lambda)$ invertible. $\square$

11.3   Two Special Local Continuation Theorems

We begin with the local version of Granas' topological transversality theorem for completely continuous maps. First we introduce the following definition. A map T from an open subset U of a Banach space X into X, is said to be *locally essential* at $x_0 \in U$ if there exists a ball $\overline{B}_r (x_0) \subset U$ such that the restriction of T to $\overline{B}_r (x_0)$ is essential in $\mathbf{M}_C \left(\overline{B}_r (x_0); X \right)$. Recall that a map belongs to $\mathbf{M}_C \left(\overline{B}_r (x_0); X \right)$ if it is compact from $\overline{B}_r (x_0)$ into X and is fixed point free on $\partial B_r (x_0)$.

Theorem 11.6 *Let Λ be a metric space, X a Banach space, $U \subset X$ open, $x_0 \in U$ and $\lambda_0 \in \Lambda$. Suppose that the following conditions are satisfied:*

 (i) $H : U \times \Lambda \to X$ is completely continuous;
 (ii) $H(x_0, \lambda_0) = x_0$;
 (iii) $H(., \lambda_0)$ is locally essential at x_0.
   Then, there exists a neighborhood V of λ_0 such that for each $\lambda \in V$, there is an $x(\lambda) \in U$ with $H(x(\lambda), \lambda) = x(\lambda)$. Moreover, if $\Lambda \subset \mathbf{R}$, then there exists a maximal interval $V \subset \Lambda$ having the following

properties: a) $\lambda_0 \in V$; *b) for each* $\lambda \in V$,   *there exists* $x(\lambda) \in U$
with $H(x(\lambda), \lambda) = x(\lambda)$ *and* $H(., \lambda)$ *is locally essential at* $x(\lambda)$. *In
addition, this maximal interval is open in* Λ.

Proof. $H(., \lambda_0)$ being locally essential at x_0 guarantees that there is
a ball $\overline{B}_r(x_0) \subset U$ such that $H(., \lambda_0)$ is essential in $\mathbf{M}_C\left(\overline{B}_r(x_0); X\right)$.
We only have to show that there is a neighborhood V of λ_0 such that
$H(., \lambda)$ is essential in $\mathbf{M}_C\left(\overline{B}_r(x_0); X\right)$ for all $\lambda \in V$. We claim
that there exists a neighborhood V of λ_0 with

$$(1 - \mu) H(x, \lambda_0) + \mu H(x, \lambda) \neq x \qquad (11.6)$$

for all $\mu \in [0, 1]$, $\lambda \in V$ and $|x - x_0| = r$. To prove this, assume the
contrary. Then, there are the sequences $\mu_k \in [0, 1]$, $\lambda_k \in \Lambda$ and x_k
with $|x_k - x_0| = r$, such that $\lambda_k \to \lambda_0$ as $k \to \infty$ and

$$(1 - \mu_k) H(x_k, \lambda_0) + \mu_k H(x_k, \lambda_k) = x_k. \qquad (11.7)$$

Clearly, we may suppose that μ_k converges to some $\mu_0 \in [0, 1]$. By the
assumption that H is completely continuous, we may also suppose that
the sequences $(H(x_k, \lambda_0))$ and $(H(x_k, \lambda_k))$ are convergent. Then, by
(11.7), (x_k) converges, say to $\bar{x}$. Letting $k \to \infty$ in (11.7), we obtain

$$(1 - \mu_0) H(\bar{x}, \lambda_0) + \mu_0 H(\bar{x}, \lambda_0) = H(\bar{x}, \lambda_0) = \bar{x},$$

where $|\bar{x} - x_0| = r$. This contradicts the property of $H(., \lambda_0)$ be-
ing fixed point free on $\partial B_r(x_0)$. This proves our claim. Next, since
$H(., \lambda_0)$ is assumed to be essential in $\mathbf{M}_C\left(\overline{B}_r(x_0); X\right)$, from (11.6)
and Theorem 5.4, we have that $H(., \lambda)$ is essential too for all $\lambda \in V$. $\square$

The following local implicit function theorem is due to Zabrejko (see
[9]) and follows from the global continuation result, Corollary 9.13.

Theorem 11.7 *Let* X *be a reflexive Banach space,* Λ *a topological
space. Let* $U_0 \subset X$ *be open with* $x_0 \in U_0$ *and* $F : U_0 \times \Lambda \to X^*$.
Suppose that there is a function $\psi : (0, \infty) \to (0, \infty)$ *such that for each*
$\lambda \in \Lambda$, $F(., \lambda)$ *is hemicontinuous and*

$$\langle F(x, \lambda) - F(y, \lambda), x - y \rangle \geq \psi(|x - y|)$$

for all $x, y \in U_0$ *with* $x \neq y$. *In addition assume that* $F(x_0, \lambda_0) = 0$
and $F(x_0, .)$ *is continuous at* λ_0 *for some* $\lambda_0 \in \Lambda$. *Then, there*

exists a neighborhood V of λ_0 and a function $x : V \to U_0$ such that $x(\lambda_0) = x_0$ and $F(x(\lambda), \lambda) = 0$ for all $\lambda \in V$.

Proof. Since any monotone hemicontinuous map from a part of a re-flexive Banach space to its dual is demicontinuous on the interior of its domain (see [118], Proposition 3.2.2), $F(.,\lambda)$ is demicontinuous on U_0. Also, according to Proposition 9.10, $F(.,\lambda)$ is pseudomonotone. We show that for each λ in some neighborhood of λ_0, Corollary 9.13 applies to $f(.,\lambda)$ given by $f(x,\lambda) = F(x + x_0, \lambda)$, with $T = 0$ and $U = B_r(0)$, where $r > 0$ is such that $\overline{B}_r(x_0) \subset U_0$. Indeed, for $|x| = r$, we have

$$\langle f(x,\lambda), x \rangle = \langle F(x + x_0, \lambda) - F(x_0, \lambda), x \rangle + \langle F(x_0, \lambda), x \rangle$$

$$\geq \psi(|x|) - |x| |F(x_0,\lambda)| = \psi(r) - r |F(x_0,\lambda)|.$$

Since $\psi(r) > 0$, $F(x_0, \lambda_0) = 0$ and $F(x_0, .)$ is continuous, there exists a neighborhood V of λ_0 with

$$\psi(r) - r |F(x_0,\lambda)| \geq 0 \quad \text{for all } \lambda \in V.$$

From Corollary 9.13, there exists $\bar{x}(\lambda) \in \overline{B}_r(0)$ with $f(\bar{x}(\lambda), \lambda) = 0$. Then,

$$x(\lambda) := \bar{x}(\lambda) + x_0 \in U_0 \quad \text{and} \quad F(x(\lambda), \lambda) = 0.$$

□

For results of the above type see [7] and the references therein. Lo-cal implicit function theorems for accretive maps derived from global continuation results appear in [34]. A good survey on local continua-tion theorems containing much literature is the recent paper by Appell, Vignoli and Zabrejko [9].

11.4 Continuation and Stability

The goal of this section is to focus on the idea that "stability" implies continuation. In fact, conditions like as: "H_0 is a constant map" (The-orem 5.3), "H_0 is essential" (Theorem 5.4), or "$\langle f(x), x \rangle \geq 0$ on ∂U" (Corollary 9.13) are all expressions of a stability-like property of the starting solution of the corresponding equation $\lambda = 0$.

We conclude with an example illustrating the above assertion and also the applicability of the local implicit function theorems. We shall discuss the existence of a maximal continuum of stable positive solutions u_λ, $0 < \lambda < \lambda^* \leq \infty$, of the one-parameter family of $(p, n-p)$ focal boundary value problems

$$\begin{cases} (-1)^{n-p} u^{(n)} = \lambda f(u), & 0 < t < 1 \\ u^{(j)}(0) = 0, & 0 \leq j \leq p-1 \\ u^{(j)}(1) = 0, & p \leq j \leq n-1 \end{cases} \tag{11.8}$$

where $1 \leq p \leq n-1$, $n \geq 2$ and $\lambda \in \mathbf{R}$.

Let $C_{\mathcal{B}}^n$ be the space of all functions $u \in C^n[0,1]$ satisfying

$$u^{(j)}(0) = 0, \quad 0 \leq j \leq p-1, \quad u^{(j)}(1) = 0, \quad p \leq j \leq n-1,$$

and let $C = C[0,1]$.

Theorem 11.8 *Suppose $f \in C^1(\mathbf{R}_+; (0, \infty))$. Then there exists a maximal $\lambda^* \in (0, \infty]$ and a C^1 function $\lambda \longmapsto u_\lambda$ from $[0, \lambda^*)$ into $C_{\mathcal{B}}^n$ such that $u_0 = 0$ and for each $\lambda \in (0, \lambda^*)$, $u_\lambda > 0$ on $(0,1)$, solves (11.8) and the map*

$$L_0 - \lambda f'(u_\lambda) : C_{\mathcal{B}}^n \to C,$$

$$(L_0 - \lambda f'(u_\lambda)) v = (-1)^{n-p} v^{(n)} - \lambda f'(u_\lambda) v$$

is invertible.

Proof. We shall apply Corollary 11.5. Let $X = \left(C_{\mathcal{B}}^n, \| \cdot \|_{n,\infty} \right)$, $\Lambda = \mathbf{R}$, $Y = (C, \| \cdot \|_\infty)$ and let $F : C_{\mathcal{B}}^n \times \mathbf{R} \to C$ be given by

$$F(u, \lambda) = (-1)^{n-p} u^{(n)} - \lambda \tilde{f}(u),$$

where $\tilde{f} \in C^1(\mathbf{R}; (0, \infty))$ is any positive smooth extension of f from $\mathbf{R}_+$ to $\mathbf{R}$.

It is easy to show that $F \in C^1(X \times \mathbf{R}; Y)$ and

$$F_u'(u, \lambda) v = (-1)^{n-p} v^{(n)} - \lambda \tilde{f}'(u) v$$

(Recall that $F_u'(u, \lambda)$ is a linear continuous map from $C_{\mathcal{B}}^n$ into C). Also,

$$F(0, 0) = 0 \quad \text{and} \quad F_u'(0, 0) v = (-1)^{n-p} v^{(n)}.$$

Now according to the preliminaries in Section 10.2, $F_u'(0, 0)$ is invertible. Thus, Corollary 11.5 implies the existence of a maximal (open) real interval V containing 0 and of a C^1 function $\lambda \longmapsto u_\lambda$ from V into C_B^n such that $u_0 = 0$ and for each $\lambda \in V$, $F(u_\lambda, \lambda) = 0$ and $F_u'(u_\lambda, \lambda)$ is invertible.

Notice for $\lambda \in V$, $\lambda > 0$,

$$(-1)^{n-p} u_\lambda^{(n)}(t) = \lambda \tilde{f}(u_\lambda(t)) > 0 \quad \text{for } t \in (0,1).$$

Now, Theorem 10.13 guarantees $u_\lambda(t) > 0$ for $t \in (0,1)$. Hence, for $\lambda \in V$, $\lambda > 0$, u_λ solves (11.8) and the extension $\tilde{f}$ of f does not play a role. Finally, let $[0, \lambda^*) = V \cap [0, \infty)$. $\square$

Notice from Theorem 10.13, we have more than the positivity of u_λ, namely

$$u_\lambda^{(j)}(t) \geq t^{p-j} \left\| u_\lambda^{(j)} \right\|_\infty > 0 \quad \text{for } t \in (0,1),$$

$j = 0, 1, ..., p-1$ and $0 < \lambda < \lambda^*$.

In what follows, we are interested in the right bound λ^* and in extra properties of the function $\lambda \longmapsto u_\lambda$, $0 \leq \lambda < \lambda^*$. For this, we shall restrict ourselves to the particular "autoadjoint" case, when $n = 2m$ $(m \geq 1)$ and $p = m$. Thus, we shall discuss the problem

$$\begin{cases} (-1)^m u^{(2m)} = \lambda f(u), & 0 < t < 1 \\ u^{(j)}(0) = 0, & 0 \leq j \leq m-1 \\ u^{(j)}(1) = 0, & m \leq j \leq 2m-1. \end{cases} \tag{11.9}$$

First, we need some notions. Let $u \in C_B^{2m}$ be a solution to

$$\begin{cases} (-1)^m u^{(2m)} + au = f, & 0 < t < 1 \\ u^{(j)}(0) = 0, & 0 \leq j \leq m-1 \\ u^{(j)}(1) = 0, & m \leq j \leq 2m-1. \end{cases} \tag{11.10}$$

where $a, f \in C$. Then, if we multiply the equation in (11.10) by any function $v \in C_B^{2m}$ and integrate, we obtain

$$\int_0^1 \left(u^{(m)} v^{(m)} + auv - fv \right) dt = 0. \tag{11.11}$$

Thus, it is natural to define a *weak solution* of (11.10) (for $a \in L^\infty(0,1)$ and $f \in L^2(0,1)$) to be a function $u \in W_B^{m,2}$ which satisfies (11.11) for all $v \in C_B^{2m}$; here

$$W_B^{m,2} = \left\{ u \in W^{m,2}(I; \mathbf{R}); \ u^{(j)}(0) = 0, \ 0 \leq j \leq m-1 \right\}.$$

The space $W_B^{m,2}$ may be endowed with the inner product and the corresponding norm

$$\langle u, v \rangle = \langle u^{(m)}, v^{(m)} \rangle_2, \quad \|u\| = \|u^{(m)}\|_2.$$

The equivalence of $\|.\|$ and $\|.\|_{m,2}$ on $W_B^{m,2}$ is immediate. Notice that if $a, f \in C$ and $u \in W_B^{m,2} \cap C^{2m}$ is a weak solution of (11.10), then u also satisfies $u^{(j)}(1) = 0$, $m \leq j \leq 2m - 1$, that is u is a classical solution of (11.10). This means that the conditions $u^{(j)}(1) = 0$, $m \leq j \leq 2m - 1$ are implied by the variational identity (11.11). Hence they are so called *natural boundary conditions*.

Next, let

$$\lambda_1(L_0 + a) = \inf \left\{ \int_0^1 \left[\left(u^{(m)}\right)^2 + au^2 \right] dt \,/\, \|u\|_2^2 ;\ u \in W_B^{m,2},\ u \neq 0 \right\}.$$
$$(11.12)$$

Lemma 11.9 *For each $a \in C$, $\lambda_1(L_0 + a)$ is the smallest (first) eigenvalue of $(L_0 + a, \mathcal{B})$ and it is continuous in a.*

Proof. Let $\lambda = \lambda_1(L_0 + a)$. We shall prove that there exists a non zero solution $\varphi \in W_B^{m,2}$ to

$$(-1)^m \varphi^{(2m)} + a\varphi = \lambda\varphi. \qquad (11.13)$$

For this, let (u_k) be any sequence with the following properties:

$$u_k \in W_B^{m,2}, \quad \|u_k\|_2 = 1 \quad \text{and}$$

$$\int_0^1 \left[\left(u_k^{(m)}\right)^2 + au_k^2 \right] dt \to \lambda \quad \text{as } k \to \infty. \qquad (11.14)$$

Hence (u_k) is bounded in $W_B^{m,2}$. Since the imbedding of $W^{m,2}$ into $L^2(0,1)$ is completely continuous, we may suppose passing if necessary to a subsequence, that $u_k \to \varphi$ in $L^2(0,1)$, where $\varphi \in L^2(0,1)$. Clearly, $\|\varphi\|_2 = 1$ and

$$\int_0^1 au_k^2 dt \to \int_0^1 a\varphi^2 dt \quad \text{as } k \to \infty. \qquad (11.15)$$

Furthermore, using the identity

$$\left\| u_k^{(m)} - u_l^{(m)} \right\|_2^2 + \left\| u_k^{(m)} + u_l^{(m)} \right\|_2^2 = 2 \left(\left\| u_k^{(m)} \right\|_2^2 + \left\| u_l^{(m)} \right\|_2^2 \right),$$

(11.14) and (11.15), we obtain

$$\left\| u_k^{(m)} - u_l^{(m)} \right\|_2^2 \le 2 \left(\left\| u_k^{(m)} \right\|_2^2 + \left\| u_l^{(m)} \right\|_2^2 \right) - \lambda \left\| u_k + u_l \right\|_2^2$$

$$+ \int_0^1 a \left(u_k + u_l \right)^2 dt \ \to \ 0 \quad \text{as} \quad k, l \to \infty.$$

This implies that $u_k \to \varphi$ in $W_B^{m,2}$. Now (11.14) guarantees

$$\lambda = \int_0^1 \left[\left(\varphi^{(m)} \right)^2 + a\varphi^2 \right] dt,$$

that is the infimum in (11.12) is attained.

Now we claim that φ is a weak solution to (11.13). Indeed, for any fixed $v \in C_B^{2m}$, the function

$$g\left(\tau \right) = \int_0^1 \left[\left((\varphi + \tau v)^{(m)} \right)^2 + a \left(\varphi + \tau v \right)^2 \right] dt \ / \ \left\| \varphi + \tau v \right\|_2^2$$

is well defined in some neighborhood of the origin and attains its infimum at $\tau = 0$. Direct computation yields

$$g'\left(0 \right) = 2 \int_0^1 \left(\varphi^{(m)} v^{(m)} + a\varphi v - \lambda \varphi v \right) dt.$$

This proves our claim.

The fact that $\lambda_1 \left(L_0 + a \right)$ is the smallest eigenvalue and its continuous dependance on a are left to the reader. $\square$

Notice (11.12) guarantees that

$$\lambda_1 \left(L_0 + a \right) > 0 \quad \text{for every} \quad a \ge 0.$$

Also, observe that

if $\lambda_1 \left(L_0 + a \right) = 0,$ then $L_0 + a$ is not invertible

and

if $\lambda_1 \left(L_0 + a \right) > 0,$ then $L_0 + a$ is *coercive*,

i.e. there exists a constant $c > 0$ with

$$\int_0^1 \left[\left(u^{(m)} \right)^2 + au^2 \right] dt \ge c \left\| u \right\|_2^2$$

for all $u \in W_B^{m,2}$ (take $c = \lambda_1 \left(L_0 + a \right)$)

For a coercive map $L_0 + a$ we have the following maximum principle.

Lemma 11.10 *Suppose that $a \in C$ and the map $L_0 + a$ is coercive. If $u \in C_B^{2m}$ is the solution of the problem (11.10) for some $f \in C$ and $f(t) \geq 0$ for all $t \in (0,1)$, then $u(t) \geq 0$ on $(0,1)$.*

The proof of this lemma follows from standard arguments and is omitted.

Theorem 11.11 *Suppose $f \in C^1(\mathbf{R}_+; (0, \infty))$ is strictly increasing and convex. Then there exists a maximal $\lambda^* \in (0, \infty)$ and a C^1 function $\lambda \longmapsto u_\lambda$ from $[0, \lambda^*)$ into C_B^{2m} such that $u_0 = 0$ and for each $\lambda \in (0, \lambda^*)$, $u_\lambda > 0$ on $(0,1)$, solves (11.9) and is stable in the sense that*

$$\lambda_1 (L_0 - \lambda f'(u_\lambda)) > 0. \tag{11.16}$$

Moreover,

$$\begin{cases} \left(\frac{du_\lambda}{d\lambda}\right)^{(j)} > 0 \text{ on } (0,1), \ 0 \leq j \leq m \\ (-1)^{j-m} \left(\frac{du_\lambda}{d\lambda}\right)^{(j)} > 0 \text{ on } (0,1), \ m+1 \leq j \leq 2m. \end{cases} \tag{11.17}$$

Proof. According to Theorem 11.8, for each $0 < \lambda < \lambda^*$, the map $L_0 - \lambda f'(u_\lambda)$ is invertible. Suppose that

$$\lambda_1 \left(L_0 - \overline{\lambda} f'(u_{\overline{\lambda}})\right) \leq 0$$

for some $\overline{\lambda} \in (0, \lambda^*)$. Then, since $\lambda_1(L_0) > 0$ and $\lambda_1(L_0 + a)$ is continuous in a, there is an $\lambda \in (0, \overline{\lambda}]$ with

$$\lambda_1 (L_0 - \lambda f'(u_\lambda)) = 0.$$

It follows that the map $L_0 - \lambda f'(u_\lambda)$ is not invertible, a contradiction. Thus, (11.16) holds for all $\lambda \in [0, \lambda^*)$.

Now we shall prove that $\lambda^* < \infty$. Suppose $\lambda^* = \infty$. Then, since f is increasing,

$$L_0 u_\lambda = \lambda f(u_\lambda) \geq \lambda f(0) \text{ on } (0,1). \tag{11.18}$$

Let $w \in C_B^{2m}$ be the solution of the problem

$$\begin{cases} (-1)^m w^{(2m)} = 1, \ 0 < t < 1 \\ w^{(j)}(0) = 0, \ 0 \leq j \leq m-1 \\ w^{(j)}(1) = 0, \ m \leq j \leq 2m-1. \end{cases} \tag{11.19}$$

From Theorem 10.13, $w > 0$ on $(0,1)$. From (11.18) and (11.19), we have

$$L_0 \left(u_\lambda - \lambda f(0) w \right) \geq 0 \quad \text{on} \quad (0,1).$$

Now Lemma 11.10 implies

$$u_\lambda(t) \geq \lambda f(0) w(t) \quad \text{for all} \quad t \in (0,1). \tag{11.20}$$

Taking into account (11.20), the increasing monotonicity of f' (since f is convex) and the coercivity of $L_0 - \lambda f'(u_\lambda)$, we obtain

$$\int_0^1 f'(\lambda_0 f(0) w) v^2 \, dt \leq \int_0^1 f'(\lambda f(0) w) v^2 \, dt$$

$$\leq \int_0^1 f'(u_\lambda) v^2 \, dt \leq \frac{1}{\lambda} \int_0^1 \left(v^{(m)} \right)^2 dt$$

for all $v \in C_B^{2m}$ and $0 < \lambda_0 \leq \lambda$. If we let $\lambda \to \infty$, this gives

$$\int_0^1 f'(\lambda_0 f(0) w) v^2 \, dt = 0 \quad \text{for all} \quad v \in C_B^{2m}.$$

Consequently,

$$f'(\lambda_0 f(0) w(t)) = 0 \quad \text{for all} \quad t \in (0,1).$$

Now fix any $t \in (0,1)$ and let $\lambda_0 \to \infty$ to deduce that $f'(s) \to 0$ as $s \to \infty$, a contradiction. Thus, $\lambda^* < \infty$.

The last step is to prove (11.17). From Theorem 11.8, we know that the map $\lambda \longmapsto u_\lambda$ is C^1. Now, if we differentiate (11.9) with respect to λ, we obtain

$$\begin{cases} (-1)^m \left(\frac{du_\lambda}{d\lambda} \right)^{(2m)} - \lambda f'(u_\lambda) \frac{du_\lambda}{d\lambda} = f(u_\lambda), \quad 0 < t < 1 \\ \left(\frac{du_\lambda}{d\lambda} \right)^{(j)}(0) = 0, \quad 0 \leq j \leq m - 1 \\ \left(\frac{du_\lambda}{d\lambda} \right)^{(j)}(1) = 0, \quad m \leq j \leq 2m - 1. \end{cases} \tag{11.21}$$

Since $L_0 - \lambda f'(u_\lambda)$ is coercive and $f(u_\lambda) \geq 0$, Lemma 11.10 guarantees that $du_\lambda / d\lambda \geq 0$. Then, since $f'(u_\lambda) \geq 0$ and $f(u_\lambda) > 0$ on $(0,1)$, we have

$$\lambda f'(u_\lambda) \frac{du_\lambda}{d\lambda} + f(u_\lambda) > 0 \quad \text{on} \quad (0,1).$$

Now (11.17) follows from (11.21) by successive integration. □

For other aspects concerning the connection between stability and continuation (degree theory) we refer the reader to Ortega [116] and the references therein.

Epilogue

In this book we have presented continuation theorems for several classes of nonlinear maps. Moreover we have placed these theorems in a global setting. Also we have discussed typical applications in differential equations.

Hopefully after reading this book the reader will be aware of the powerful applicability of these theorems. As a result the reader may be interested in establishing new such continuation theorems or in applying our results to other problems.

1) If the reader is mainly interested in abstract results, he or she should remember two basic ideas in establishing continuation theorems. Suppose one wishes to discuss the solvability of the inclusion $\Gamma_1(x) \in B \subset \Theta$, $x \in \Xi$. The basic idea in this book is to try to deduce it from a simpler inclusion $\Gamma_0(x) \in B$, by continuation, i.e. by using a suitable homotopy $\eta : \Xi \times [0,1] \to \Theta$ joining Γ_0 and Γ_1. The reader could try

(a) to prove that the set $\Lambda_\eta = \{\lambda \in [0,1] \; ; \; \eta(x,\lambda) \in B$ for some $\lambda \in [0,1]\}$ is open and closed simultaneously in $[0,1]$,

or

(b) assuming that Ξ is the closure of an open subset U of a metric space, to prove that the set $\Sigma_\eta = \{x \in \overline{U}; \; \eta(x,\lambda) \in B$ for some $\lambda \in [0,1]\}$ is closed and disjoint of ∂U, and then that for each map of the form $\eta(\,.\,,\upsilon(\,.\,))$, where $\upsilon \in C\left(\overline{U}; [0,1]\right)$, $\upsilon(x) = 0$ on ∂U, $\upsilon(x) = 1$ on Σ_η, there exists at least one solution to the inclusion $\eta(x,\upsilon(x)) \in B$.

Also the reader should be aware that the time of broad classes of maps has gone. Thus, instead of dreaming up continuation theorems for some larger class of maps, it may be more useful to look at particular "pathological" maps which arise in mathematical modelling.

In regards to the topics in this book, we stress in particular the selective continuation results. In particular we mention nonlinear problems without a priori bounds on solutions. There are many unanswered questions in this area and we hope many mathematicians, young and old, will become interested in these problems.

Another exciting direction concerns the interaction between topological methods in general (the continuation ones in particular) and other kinds of methods, for example variational methods, numerical methods, or methods from adjacent fields such as algebraic topology.

2) If the reader is interested more in applications, then he or she will need to discover and exploit the properties of the specific operator corresponding to each particular problem and so to identify the abstract result which is directly applicable.

One of the main reasons for writing this book was to present abstract results with connections to concrete applications. This was not only our personal goal but also a basic principle for most applied mathematics today. Therefore, in our opinion, the correct way to study and present research in nonlinear analysis is to go from applications to theory and back again to applications.

Bibliography

[1] R.P. Agarwal, *Boundary Value Problems for Higher Order Differential Equations*, World Scientific, Singapore, 1986.

[2] R.P. Agarwal, *Focal Boundary Value Problems for Differential and Difference Equations*, Kluwer, Dordrecht, 1998.

[3] R.P. Agarwal and D. O'Regan, *Nonlinear superlinear singular and nonsingular second order boundary value problems*, J. Differential Equations **143** (1998), 60-95.

[4] R.P. Agarwal, D. O'Regan and V. Lakshmikantham, *A note on multiple solutions for (p,n-p) focal and (n,p) problems*, Nonlinear Stud., to appear.

[5] R.P. Agarwal and P.J.Y. Wong, *Error Inequalities in Polynomial Interpolation and Their Applications*, Kluwer, Dordrecht, 1993.

[6] R.R. Akhmerov, M.I. Kamenskii, A.S. Potapov, A.E. Rodkina and B.N. Sadovskii, *Measures of Noncompactness and Condensing Operators* (Russian), Nauka, Novosibirsk, 1986; Engl. transl.: Birkhäuser, Basel, 1992.

[7] W. Alt and I. Kolumbán, *Implicit-function theorems for monotone mappings*, Kybernetika **29** (1993), 210-221.

[8] A. Ambrosetti, *Un teorema di esistenza per le equazioni differenziali negli spazi di Banach*, Rend. Sem. Mat. Univ. Padova **39** (1967), 349-360.

[9] J. Appell, A. Vignoli and P.P. Zabrejko, *Implicit function theorems and nonlinear integral equations*, Expo. Math. **14** (1996), 385-424.

[10] D.D. Bainov and S.I. Kostadinov, *Abstract Impulsive Differential Equations,* Descartes Press, Koriyama, 1995.

[11] J. Banas and K. Goebel, *Measures of Noncompactness in Banach Spaces,* Marcel Dekker, New York, 1980.

[12] V. Barbu, *Nonlinear Semigroups and Differential Equations in Banach Spaces,* Ed. Academiei-Noordhoff, Bucureşti, Leyden, 1976.

[13] P.R. Beesack and K.M. Das, *Extensions of Opial's inequality,* Pacific J. Math. **26** (1968), 215-231.

[14] H. Ben-El-Mechaiekh and G. Isac, *Some geometric solvability theorems in topological vector spaces,* Bull. Korean Math. Soc. **34** (1997), 273-285.

[15] S. Bernstein, *Sur les équations du calcul des variations,* Ann. Sci. Ecole Norm. Sup. **29** (1912), 431-485.

[16] L.E. Bobisud, *The desintegrating bar: reaction-diffusion in a moving medium,* Nonlinear Anal. **21** (1993), 131-141.

[17] H. Brezis, *Analyse fonctionnelle,* Masson, Paris, 1983.

[18] F.E. Browder, *Nonexpansive nonlinear operators in a Banach space,* Proc. Nat. Acad. Sci. U.S.A. **54** (1965), 1041-1044.

[19] F.E. Browder, *Nonlinear operators and nonlinear equations of evolution in Banach spaces,* Proc. Sympos. Nonlinear Functional Analysis, Amer. Math. Soc, Chicago, 1968.

[20] F.E. Browder, *Fixed point theory and nonlinear problems,* Bull. Amer. Math. Soc. **9** (1983), 1-39.

[21] A. Cabada and J.J. Nieto, *Approximation of solutions for nonlinear second-order boundary value problems,* Acad. Roy. Belg. Bull. Cl. Sci. (6) **2** (1991), no. 10-11, 287-311.

[22] A. Capietto, M. Henrard, J. Mawhin and F. Zanolin, *A continuation approach to some forced superlinear Sturm-Liouville boundary value problems,* Topol. Methods Nonlinear Anal. **3** (1994), 81-100.

[23] A. Capietto, J. Mawhin and F. Zanolin, *A continuation approach to superlinear periodic boundary value problems*, J. Differential Equations **88** (1990), 347-395.

[24] A. Capietto, J. Mawhin and F. Zanolin, *Continuation theorems for periodic perturbations of autonomous systems*, Trans. Amer. Math. Soc. **329** (1992), 41-72.

[25] A. Castro, *A semilinear Dirichlet problem*, Canad. J. Math. **31** (1979), 337-340.

[26] A. Constantin, *Topological transversality: application to an integrodifferential equation*, J. Math. Anal. Appl. **197** (1996), 855-863.

[27] C. Corduneanu, *Integral Equations and Applications*, Cambridge Univ. Press, New York, 1991.

[28] R. Dautray and J.L. Lions, *Analyse mathématique et calcul numérique pour les sciences et les techniques*, Masson, Paris, 1987.

[29] D.G. De Figueiredo, *The Dirichlet Problem for Nonlinear Elliptic Equations: A Hilbert Space Approach*, Lecture Notes in Math. 446, Springer, Berlin, 1974.

[30] K. Deimling, *Zeros of accretive operators*, Manuscripta Math. **13** (1974), 365-374.

[31] K. Deimling, *Ordinary Differential Equations in Banach Spaces*, Lecture Notes in Math. 596, Springer, Berlin, 1977.

[32] K. Deimling, *Nonlinear Functional Analysis*, Springer, Berlin, 1985.

[33] J. Diestel, *Geometry of Banach Spaces*, Lecture Notes in Math. 485, Springer, Berlin, 1975.

[34] A. Domokos, *Implicit function theorems for m-accretive and locally accretive set-valued mappings*, Nonlinear Anal., to appear.

[35] J. Dugundji, *Topology*, Allyn and Bacon, Boston, 1966.

[36] J. Dugundji and A. Granas, *Fixed Point Theory I*, Monografie Mat. 61, Polish Sci. Publishers, Warszawa, 1982.

[37] N. El Khattabi, *Problèmes périodiques du second ordre à croissance au plus linéaire*, Topol. Methods Nonlinear Anal. **5** (1995), 365-383.

[38] L. Erbe and W. Krawcewicz, *Existence of solutions to boundary value problems for impulsive second order differential inclusions*, Rocky Mountain J. Math. **22** (1992), 519-540.

[39] W. Feng and J.R.L. Webb, *Solvability of m-point boundary value problems with nonlinear growth*, J. Math. Anal. Appl. **212** (1997), 467-480.

[40] P.M. Fitzpatrick, *Existence results for equations involving noncompact perturbations of Fredholm mappings with applications to differential equations*, J. Math. Anal. Appl. **66** (1978), 151-177.

[41] P.M. Fitzpatrick, I. Massabò and J. Pejsachowicz, *Complementing maps, continuation and global bifurcation*, Bull. Amer. Math. Soc. **9** (1983), 79-81.

[42] A. Fonda and F. Zanolin, *Bounded solutions of nonlinear second order ordinary differential equations*, Discrete Contin. Dynam. Systems **4** (1998), 91-98.

[43] M. Frigon, *Application de la théorie de la transversalité topologique à des problèmes non linéaires pour des équations différentielles ordinaires*, Dissertationes Math. 296 (1990).

[44] M. Frigon and A. Granas, *Résultats du type de Leray-Schauder pour des contractions multivoques*, Topol. Methods Nonlinear Anal. **4** (1994), 197-208.

[45] M. Frigon, A, Granas and Z. Guennoun, *Alternative non-linéaire pour les applications contractantes*, Ann. Sci. Math. Québec. **19** (1995), 65-68.

[46] M. Frigon and J.W. Lee, *Existence principles for Carathéodory differential equations in Banach spaces*, Topol. Methods Nonlinear Anal. **1** (1993), 95-111.

[47] M. Furi, M. Martelli and A. Vignoli, *On the solvability of nonlinear operator equations in normed spaces*, Ann. Mat. Pure Appl. **124** (1980), 321-343.

[48] M. Furi and M.P. Pera, *On the existence of an unbounded set of solutions for nonlinear equations in Banach spaces*, Atti Accad. Naz. Lincei Rend. Cl. Sci. Fis. Mat. Natur. (8) **67** (1979), 31-38.

[49] R.E. Gaines and J. Mawhin, *Coincidence Degree and Nonlinear Differential Equations*, Lecture Notes in Math. 568, Springer, Berlin, 1977.

[50] J.A. Gatica and W.A. Kirk, *Fixed point theorems for contraction mappings with applications to nonexpansive and pseudo-contractive mappings*, Rocky Mountain J. Math. **4** (1974), 69-79.

[51] K. Geba, A. Granas, T. Kaczynski and W. Krawcewicz, *Homotopie et équations non linéaires dans les espaces de Banach*, C. R. Acad. Sci. Paris **300** (1985).

[52] D. Gilbarg and N. Trudinger, *Elliptic Partial Differential Equations of Second Order*, Springer, Berlin, 1983.

[53] K. Göebel, *An elementary proof of the fixed point theorem of Browder and Kirk*, Michigan Math. J. **16** (1969), 381-383.

[54] D. Göhde, *Zum prinzip der kontractiven abbildung*, Math. Narchr. **30** (1965), 251-258.

[55] A. Granas, *Homotopy extension theorem in Banach spaces and some of its applications to the theory of non-linear equations*, Bull. Acad. Pol. Sci. **7** (1959), 387-394.

[56] A. Granas, *Continuation method for contractive maps*, Topol. Methods Nonlinear Anal. **3** (1994), 375-379.

[57] A. Granas, R.B. Guenther and J.W. Lee, *Nonlinear boundary value problems for ordinary differential equations*, Dissertationes Math. 244 (1985).

[58] A. Granas, R.B. Guenther and J.W. Lee, *Some general existence principles in the Carathéodory theory of nonlinear differential systems*, J. Math. Pures Appl. **70** (1991), 153-196.

[59] R.B. Guenther and J.W. Lee, *Some existence results for nonlinear integral equations via topological transversality*, J. Integral Eqn. Appl. **5** (1993), 195-209.

[60] C.P. Gupta and S. Trofimchuk, *Existence of a solution of a three-point boundary value problem and the spectral radius of a related linear operator*, Nonlinear Anal. **34** (1998), 489-507.

[61] G.B. Gustafson and K. Schmitt, *Nonzero solution of BVP's for second order ordinary and delay differential equations*, J. Differential Equations **12** (1972), 129-147.

[62] D.D. Hai and K. Schmitt, *Existence and uniqueness results for nonlinear boundary value problems*, Rocky Mountain J. Math. **24** (1994), 77-91.

[63] P. Hartman, *Ordinary Differential Equations*, Wiley, New York, 1964.

[64] D.H. Hyers, G. Isac and Th. M. Rassias, *Topics in Nonlinear Analysis & Applications*, World Scientific, Singapore, 1997.

[65] G. Isac, *Branches continues de vecteurs propres généralisés. Applications aux équations de coïncidences*, Studia Sci. Math. Hungar. **20** (1985), 155-172.

[66] G. Isac, *On an Altman type fixed point theorem on convex cones*, Rocky Mountain J. Math. **25** (1995), 701-714.

[67] G. Isac, *On the solvability of multivalued complementarity problem: a topological method*, to appear.

[68] T. Kaczynski and J. Wu, *A topological transversality theorem for multi-valued maps in locally convex spaces with applications to neutral equations*, Can. J. Math. **44** (5) (1992), 1003-1013.

[69] R. Kannan and R. Ortega, *Periodic solutions of pendulum-type equations*, J. Differential Equations **59** (1985), 123-144.

[70] J.L. Kazdan and F.W. Warner, *Remarks on some quasilinear elliptic equations*, Comm. Pure Appl. Math. **28** (1975), 567-597.

[71] W.A. Kirk, *A fixed point theorem for mapping which do not increase distance*, Amer. Math. Monthly **72** (1965), 96-98.

[72] W.A. Kirk and R. Schöneberg, *Some results on pseudo-contractive mappings*, Pacific J. Math. **71** (1977), 89-100.

[73] E. Kirr and R. Precup, *Periodic solutions of superlinear impulsive differential systems*, Comm. Appl. Anal., **3** (1999), 483-502.

[74] M.A. Krasnoselskii, *Topological Methods in the Theory of Nonlinear Integral Equations* (Russian), Gostekhteoretizdat, Moscow, 1956; Engl. transl.: Pergamon Press, Oxford, 1964.

[75] M.A. Krasnoselskii and P.P. Zabrejko, *Geometrical Methods of Nonlinear Analysis* (Russian), Nauka, Moskow, 1975; Engl. transl.: Springer, Berlin, 1984.

[76] W. Krawcewicz, *Contribution à la théorie des équations non-linéaires dans les espaces de Banach*, Dissertationes Math. 273 (1988).

[77] W. Krawcewicz, *Résolution des équations semilinéaires avec la partie linéaire à noyau de dimension infinie via des applicationes A-propres*, Dissertationes Math. 294 (1990).

[78] V. Lakshmikantham and S. Leela, *Nonlinear Differential Equations in Abstract Spaces*, Pergamon Press, New York, 1981.

[79] A. Lasota and Y.A. Yorke, *Existence of solutions of two point boundary value problems for nonlinear systems*, J. Differential Equations **11** (1972), 509-518.

[80] J.W. Lee and D. O'Regan, *Nonlinear boundary value problems in Hilbert spaces*, J. Math. Anal. Appl. **137** (1989), 59-69.

[81] J.W. Lee and D. O'Regan, *Existence results for differential equations in Banach spaces*, Comment. Math. Univ. Carolin. **34** (1993), 239-252.

[82] J.W. Lee and D. O'Regan, *Existences principles for differential equations and systems of equations*, in: Topological Methods in Differential Equations and Inclusions, A. Granas and M. Frigon (eds.), Kluwer, Dordrecht, 1995, 239-289.

[83] J. Leray and J. Schauder, *Topologie et équations fonctionnelles*, Ann. Sci. École Norm. Sup. **51** (1934), 45-78.

[84] N.G. Lloyd, *Degree Theory*, Cambridge Univ. Press, Cambridge, 1978.

[85] M.G. Maia, *Un'obsservazione sulle contrazioni metriche*, Rend. Sem. Mat. Univ. Padova **40** (1968), 139-143.

[86] M. Martelli, *Continuation Principles and Boundary Value Problems*, Lecture Notes in Math. 1537, Springer, Berlin, 1993.

[87] R.H. Martin, *Nonlinear Operators and Differential Equations in Banach Spaces*, Wiley, New York, 1976.

[88] J. Mawhin, *Compacité, monotonie et convexité dans l'étude de problèmes aux limites semi-linéaires*, Séminaire d'analyse moderne 19, Université de Sherbrooke, 1981.

[89] J. Mawhin, *Continuation theorems and periodic solutions of ordinary differential equations*, in: Topological Methods in Differential Equations and Inclusions, A. Granas and M. Frigon (eds.), Kluwer, Dortdrecht, 1995, 291-375.

[90] J. Mawhin and K. Rybakovski, *Continuation theorems for semilinear equations in Banach spaces: A survey*, in: Nonlinear Analysis, Th. Rassias (ed.), World Scientific, Singapore, 1987, 367-405.

[91] J. Mawhin and J. Ward Jr., *Nonresonance and existence for nonlinear elliptic boundary value problems*, Nonlinear Anal. **6** (1981), 677-684.

[92] P.S. Milojevic, *Approximation-solvability of nonlinear equations and applications*, in: Fourier Analysis: Analytic and Geometric Aspects, W.O. Bray et al (eds.), Lecture Notes in Pure Appl. Math. 157, Marcel Dekker, New York, 1994, 311-375.

[93] H. Mönch, *Boundary value problems for nonlinear ordinary differential equations of second order in Banach spaces*, Nonlinear Anal. **4** (1980), 985-999.

[94] H. Mönch and G.F. von Harten, *On the Cauchy problem for ordinary differential equations in Banach spaces*, Arch. Math. (Basel) **39** (1982), 153-160.

[95] C. Morales, *Pseudo-contractive mappings and the Leray-Schauder boundary condition*, Comment. Math. Univ. Carolin. **20** (1979), 745-756.

[96]   C. Morales, *Existence theorems for demicontinuous accretive operators in Banach spaces*, Houston J. Math. **10** (1984), 535-543.

[97]   J.J. Nieto, *Basic theory for nonresonance impulsive periodic problems of first order*, J. Math. Anal. Appl. **205** (1997), 423-433.

[98]   L. Nirenberg, *Variational and topological methods in nonlinear problems*, Bull. Amer. Math. Soc. (New Ser.) **4** (1981), 267-302.

[99]   S.K. Ntouyas, Y.G. Sficas and P.Ch. Tsamatos, *Boundary value problems for functional differential equations*, J. Math. Anal. Appl. **199** (1996), 213-230.

[100] S.K. Ntouyas and P.Ch. Tsamatos, *Existence of solutions of boundary value problems for differential equations with deviating arguments, via the topological transversality method*, Proc. Roy. Soc. Edinburgh Sect. A **118** (1991), 79-89.

[101] R.D. Nussbaum, *Degree theory for local condensing maps*, J. Math. Anal. Appl. **37** (1972), 741-766.

[102] D. O'Regan, *Theory of Singular Boundary Value Problems*, World Scientific, Singapore, 1994.

[103] D. O'Regan, *Measure of noncompactness, Darbo maps and differential equations in abstract spaces*, Acta Math. Hungar. **69**, no. 3 (1995), 233-261.

[104] D. O'Regan, *Fixed point theorems for nonlinear operators*, J. Math. Anal. Appl. **202** (1996), 413-432.

[105] D. O'Regan, *Nonresonant nonlinear singular problems in the limit circle case*, J. Math. Anal. Appl. **197** (1996), 708-725.

[106] D. O'Regan, *Existence of nonnegative solutions to superlinear nonpositive problems via a fixed point theorem in cones of Banach spaces*, Dynam. Contin. Discrete Impuls. Systems **3** (1997), 517-530.

[107] D. O'Regan, *Singular Dirichlet boundary value problem I: superlinear and nonresonance case*, Nonlinear Anal. **29** (1997), 221-245.

[108] D. O'Regan, *Fixed point theory for compact upper semicontinuous or lower semicontinuous maps,* Proc. Amer. Math. Soc. **125** (1997), 875-881.

[109] D. O'Regan, *A continuation theory for weakly inward maps,* Glasgow Math. J. **40** (1998), 311-321.

[110] D. O'Regan, *Fixed points for set valued mappings in locally convex linear topological spaces,* Math. Comput. Modelling **28** (1) (1998), 45-55.

[111] D. O'Regan, *Fixed point theory for closed multifunctions,* Arch. Math. (Brno) **34** (1) (1998), 191-197.

[112] D. O'Regan, *Fixed point theorems and equilibrium points in abstract economies,* Bull. Austral. Math. Soc. **58** (1998), 33-41.

[113] D. O'Regan, *Fixed points and random fixed points for weakly inward approximable maps,* Proc. Amer. Math. Soc. **126** (1998), 3045-3053.

[114] D. O'Regan, *Volterra and Urysohn integral equations in Banach spaces,* J. Appl. Math. Stochastic Anal. **11** (1998), 449-464.

[115] D. O'Regan and R. Precup, *Existence criteria for integral equations in Banach spaces,* J. Inequal. Applic.

[116] R. Ortega, *Some applications of the topological degree to stability theory,* in: Topological Methods in Differential Equations and Inclusions, A. Granas and M. Frigon (eds.), Kluwer, Dordrecht, 1995, 377-409.

[117] B.G. Pachpatte, *Applications of the Leray-Schauder alternative to some Volterra integral and integrodifferential equations,* Indian J. Pure Appl. Math. **26** (1995), 1161-1168.

[118] D. Pascali, *Continuation principles for semilinear evolution equations,* Courant Institute, preprint, 1994.

[119] D. Pascali and S. Sburlan, *Nonlinear Mappings of Monotone Type,* Ed. Academiei and Sijthoff & Noordhoff, Bucureşti, Alphen aan den Rijn, 1978.

[120] B. Petracovici, *Nonlinear two point boundary value problems*, in: Seminar on Differential Equations, Babeş-Bolyai University, Faculty of Mathematics and Physics, Research Seminars, Preprint Nr. 3, 1989, 1-12.

[121] A. Petruşel, *Fredholm-Volterra integral equations and Maia's theorem*, in: Seminar on Fixed Point Theory, Babeş-Bolyai University, Faculty of Mathematics and Physics, Research Seminars, Preprint Nr. 3, 1988, 79-82.

[122] V. Petryshyn, *Using degree theory for densely defined A-proper maps in the solvability of semilinear equations with unbounded and noninvertible linear part*, Nonlinear Anal. **4** (1980), 259-281.

[123] V. Petryshyn, *Generalized Topological Degree and Semilinear Equations*, Cambridge Univ. Press, Cambridge, 1995.

[124] H. Poincaré, *Sur certaines solutions particulières du problème des trois corps*, Bull. Astronom. **1** (1884), 65-74.

[125] H. Poincaré, *Sur un théorème de géométrie*, Rend. Circ. Mat. Palermo **33** (1912), 357-407.

[126] R. Precup, *A fixed point theorem of Maia type in syntopogenous spaces*, in: Seminar on Fixed Point Theory, Babeş-Bolyai University, Faculty of Mathematics and Physics, Research Seminars, Preprint Nr. 3, 1988, 49-70.

[127] R. Precup, *Nonlinear boundary value problems for infinite systems of second-order functional differential equations*, in: Seminar on Differential Equations, Babeş-Bolyai University, Faculty of Mathematics and Physics, Research Seminars, Preprint Nr. 8, 1988, 17-30.

[128] R. Precup, *Measure of noncompactness and second order differential equations with deviating argument*, Studia Univ. Babeş-Bolyai Math. **34** (1989), No. 2, 25-35.

[129] R. Precup, *Topological transversality, perturbation theorems and second order differential equations*, in: Seminar on Differential Equations, Babeş-Bolyai University, Faculty of Mathematics and Physics, Research Seminars, Preprint Nr. 3, 1989, 149-164.

[130] R. Precup, *Topological transversality and applications*, in: Proc. 20th Nat. Conf. Geometry and Topology Timişoara 1989, Timişoara University, 1990, 193-197.

[131] R. Precup, *Generalized topological transversality and mappings of monotone type*, Studia Univ. Babeş-Bolyai Math. **35** (1990), No. 2, 44-50.

[132] R. Precup, *Generalized topological transversality and existence theorems*, Libertas Math. **11** (1991), 65-79.

[133] R. Precup, *Topological transversality and boundary problems for second-order functional differential equations,* in: Differential Equations and Control Theory, V. Barbu (ed.), Longman, Harlow, 1991, 283-288.

[134] R. Precup, *On the topological transversality principle,* Nonlinear Anal. **20** (1993), 1-9.

[135] R. Precup, *Foundations of the continuation principles of Leray-Schauder type,* in: Proc. 23rd Conf. Geometry and Topology Cluj 1993, Cluj University, 1994, 136-140.

[136] R. Precup, *On some fixed point theorems of Deimling,* Nonlinear Anal. **23** (1994), 1315-1320.

[137] R. Precup, *Existence results for nonlinear boundary value problems under nonresonance conditions,* in: Qualitative Problems for Differential Equations and Control Theory, C. Corduneanu (ed.), World Scientific, Singapore, 1995, 263-273.

[138] R. Precup, *A Granas type approach to some continuation theorems and periodic boundary value problems with impulses,* Topol. Methods Nonlinear Anal. **5** (1995), 385-396.

[139] R. Precup, *On the continuation principle for nonexpansive maps,* Studia Univ. Babeş-Bolyai Math. **41** (1996), No. 3, 85-89.

[140] R. Precup, *Continuation theorems for maps of Caristi type,* Studia Univ. Babeş-Bolyai Math. **41** (1996), No. 4, 101-106.

[141] R. Precup, *Existence and approximation of positive fixed points of nonexpansive maps*, Rev. Anal. Numér. Théor. Approx. **26** (1997), 203-208.

[142] R. Precup, *Existence theorems for nonlinear problems by continuation methods*, Nonlinear Anal. **30** (1997), 3313-3322.

[143] R. Precup, *Continuation principles for coincidences*, Mathematica (Cluj) **39** (62) (1997), 103-110.

[144] R. Precup, *Partial Differential Equations* (Romanian), Transilvania Press, Cluj, 1997.

[145] R. Precup, *Periodic solutions of superlinear singular ordinary differential equations*, Differential Integral Equations, to appear.

[146] R. Precup, *Discrete continuation method for boundary value problems on bounded sets in Banach spaces*, J. Comput. Appl. Math. **113** (2000), 267-281.

[147] P.H. Rabinowitz, *Some aspects of nonlinear eigenvalue problems*, Rocky Mountain J. Math. **3** (1973), 161-202.

[148] S.M. Robinson, *An implicit-function theorem for a class of nonsmooth functions*, Math. Oper. Res. **16** (1991), 292-309.

[149] I.A. Rus, *On a fixed point theorem of Maia*, Studia Univ. Babeş -Bolyai Math. **22** (1977), No. 1, 40-42.

[150] I.A. Rus, *Principles and Applications of Fixed Point Theory* (Romanian), Ed. Dacia, Cluj, 1979.

[151] S. Sburlan, *Topological Degree* (Romanian), Ed. Academiei, Bucureşti, 1983.

[152] H.H. Schaefer, *Über die Methode der a-priori Schranken*, Math. Ann. **129** (1955), 415-416.

[153] K. Schmitt and R. Thompson, *Boundary value problems for infinite systems of second order differential equations*, J. Differential Equations **18** (1975), 277-295.

[154] R. Schöneberg, *Leray-Schauder principles for condensing multi-valued mappings in topological linear spaces*, Proc. Amer. Math. Soc. **72** (1978), 268-270.

[155] R. Schöneberg, *Zeros of nonlinear monotone operators in Hilbert spaces*, Canad. Math. Bull. **21** (1978), 213-219.

[156] A. Simon and P. Volkmann, *Existence of ground states with exponential decay for semi-linear elliptic equations in R^n*, J. Differential Equations **76** (1988), 374-390.

[157] A. Tiryaki, *Periodic solutions in ordinary differential equations using dissipativity conditions*, J. Math. Anal. Appl. **173** (1993), 308-317.

[158] P. Volkmann, *Démonstration d'un théorème de coïncidence par la méthode de Granas*, Bull. Soc. Math. Belg. Sér. B **36** (1984), 235-242.

[159] A. Wintner, *The nonlocal existence problem for ordinary differential equations*, Amer. J. Math. **67** (1945), 277-284.

[160] E. Zeidler, *Nonlinear Functional Analysis and its Applications*, Vol. I, Springer, Berlin, 1986.

[161] E. Zeidler, *Nonlinear Functional Analysis and its Applications*, Vol. II/B, Springer, Berlin, 1990.

Index

A
accretive map, 54
admissible map, 137, 143
Arzela-Ascoli theorem, 72

B
ball measure of noncompactness, 74
Beesack-Das inequality, 129
Bielecki norm, 22, 77
boundary value problem, 25, 61

C
Caratheodory conditions, 83
Caratheodory function, 20
Caratheodory solution, 20
Cauchy problem, 20, 74
classical solution, 20
coercive map, 185
coincidence, 101
compact map, 66
completely continuous map, 66
completion, 11
condensing map, 66
cone compression theorem, 158
cone expansion theorem, 160
connected set, 116
continuation theorem, 1
continuously differentiable map, 176
continuum, 116

contraction principle, 9

D
map of Darbo type, 66
demicoftanuous map, 54
duality map, 53

E
eigenfunctions, 85
eigenvalues, 85
elliptic problem, 83
essential map, 69, 103, 150, 3, 138, 143

F
fixed point structure, 145
focal boundary value problem, 163
Frechet derivative, 176
Frechet differentiable map, 176
Fredholm map, 101

H
Hartman inequality, 39
Hausdorff-Pompeiu metric, 71
hemicontinuous map, 148
Hilbert base, 85
homotopic maps, 156

I
implicit function, 171

Printed and bound by CPI Group (UK) Ltd, Croydon, CR0 4YY

23/10/2024

01778248-0005